T0419424

AGRICULTURE ISSUES AND POLICIES

OATS
CULTIVATION, USES AND HEALTH EFFECTS

AGRICULTURE ISSUES AND POLICIES

Additional books in this series can be found on Nova's website under the Series tab.

Additional E-books in this series can be found on Nova's website under the E-books tab.

AGRICULTURE ISSUES AND POLICIES

OATS
CULTIVATION, USES AND HEALTH EFFECTS

DANIELLE L. MURPHY
EDITOR

Nova Science Publishers, Inc.
New York

For permission to use material from this book please contact us:
Telephone 631-231-7269; Fax 631-231-8175
Web Site: http://www.novapublishers.com

Additional color graphics may be available in the e-book version of this book.

Library of Congress Cataloging-in-Publication Data

Oats : cultivation, uses and health effects / editor: Danielle L. Murphy.
p. cm.
Includes bibliographical references and index.
ISBN 978-1-61324-277-3 (hardcover : alk. paper) 1. Oats. 2. Oats as food. I. Murphy, Danielle L. II. Title.

SB191.O2O28 2011
633.1'3--dc22

2011010124

Published by Nova Science Publishers, Inc. † New York

CONTENTS

PREFACE

Oats are an ancient traditional cereal crop grown in all parts of the world and are consumed as feed and food in developing countries. Recent research as certain the health benefits and nutraceutical importance of oats in the daily diet due to the presence of dietary fiber. In this book, the authors present topical research in the study of the cultivation, uses and health effects of oats. Topics discussed in this compilation include Oat B-Glucans molecular characteristics and health benefits; the stability and degradation of soluble B-Glucan in aqueous processing; oat contamination with mycotoxins and tailoring oats for future food production.

Chapter 1 - Oat is a linear polysaccharide composed of D-glucopyranosyl residues arranged as mixtures of consecutively β-(1→4)-linked glucose units in blocks that are separated by single β-(1→3) linkages. As a soluble dietary fiber, oat has received extensive attention for its physiological effects on lowering cholesterol and postprandial blood glucose levels. The benefits are generally ascribed to the viscosity that formed in gastrointestinal tract by oat . The viscosity depends on the solubility, molecular weight, and structure of oat. So these molecular characteristics and the effect of processing on which are one of the focus of this review. The quantitative relationship between the hypoglycemic or hypocholesterolemic activities and the physicochemical characteristics of oat in food matrix will be discussed. Since cereal is an ingredient in functional food, many processes have been developed to produce concentrates and isolates, which are widely applied in food industry.

Like mushroom-derived (1→3)-, oat displays immune activity, the recent discoveries on immune function will be given a brief introduction.

Chapter 2 - The most abundant soluble dietary fibre of oat, (1→4),(1→3)-β-D-glucan, is a linear polysaccharide, which is evidently beneficial for human health. The health promoting effects are generally related to ability of beta-glucan to form highly viscous solutions, which in many foods define technological functionality of beta-glucan as well. The solution properties are influenced by parameters such as molar mass, extractability, solubility and structure of beta-glucan. Degradation of beta-glucan influences molar mass, but also its extractability and conformation in certain conditions. Food manufacturing processes cause degradation, which may be a threat for the functionality of beta-glucan if the degradation mechanisms were not understood. Degradation may as well serve tools for modification and innovation as soon as it is well-managed. However, only few degradation mechanisms are generally considered, and non-enzymatic degradation such as oxidation has been highly neglected in beta-glucan related literature until our recent publications. This review will discuss process-induced degradation of oat beta-glucan concentrating on aqueous processing. Biodegradation, chemical degradation, thermal degradation, mechanical energy induced degradation and oxidative cleavage will be discussed as degradation mechanisms of oat beta-glucan.

Chapter 3 - Oats, like other temperate cereals, can be invaded by a range of fungi both in the field before harvest and during storage. Some of these fungal species, belonging mainly to the genera *Fusarium*, *Aspergillus* and *Penicillium*, can produce a wide range of secondary metabolites. Some have the ability to contaminate the ripening grain pre-harvest or grain post-harvest with mycotoxins which can be toxic to animals and humans. Some are proven or suspected carcinogens.

Most mycotoxins are very heat stable so that once formed they are difficult to eliminate from the food supply. Cereals used for animal feed often utilise co-products that in general are likely to contain higher amounts of mycotoxins. The animal tissues will take up these chemical compounds and in this way the mycotoxins can be re-introduced into the human food chain.

Because of this, from the mid 1980's onwards, and due to the use of oat for human nutrition becoming more popular, many studies have described contamination of oats with different mycotoxin families including fusariotoxins (nivalenol, deoxynivalenol, zearalenone, HT-2 and T-2), ochratoxins and aflatoxins.

This chapter will be focused on the contamination of oats with different mycotoxins with special attention given to those produced by *Fusarium* species, the ecology of the species involved, and the most recent data on levels of contamination with Type A trichotecenes in northern European countries. The European legislation regarding mycotoxin contamination of cereals will be highlighted and, finally, reduction and control strategies will be discussed.

Chapter 4 - Oats (*Avena sativa* L.) is one of the important crops of arid and semi arid zone and is rich in polysaccharides, vitamins proteins, phenolic compounds and antioxidants compounds. Especially its component, β-glucan has got considerable importance because of its numerous industrial, nutritional and health benefits. It has a great prospective to find its application in nutraceutical food industry in this century that is driven by convenience, health, taste and sustainable production. But unfortunately, at present most of this crop is being used only as feed for animals and its actual potential is yet to be explored for industrial purposes. Different food agencies and research organizations are in the process of recognizing health benefits of oat, it is therefore timely to contemplate potential of oat crop for developing nutraceutical industry in this millennium.

Important functional ingredients in oats are dietary fiber, vitamins, proteins, phenolic compounds, antioxidants and a special nutraceutical ingredient known as β-glucan (β-1→3, 1→4 glucan). The food industry can take advantage of the unique properties of oat β-glucan, to use these for production of new food products. But extraction and purification of β-glucan involves a complex process and require special attention to capitalize on its yield and functional properties.

Previous research on β-glucan has demonstrated its multiple human health benefits. Such as its tendency to reduce onset of colorectal cancer, increased stool bulk, mitigate constipation, reduction in glycemic index, flattening of the postprandial blood glucose levels and insulin rises. Keeping in view of its health benefits, a range of functional foods containing β-glucan are now being commercially introduced to the market. Thus, the industrial demand for this natural cereal based compound is fast growing and has a great potential in future foods. This chapter contains details about significance of nutraceutical components from oat, β-glucan extraction techniques, β-glucan yield and recovery, chemical analysis of β-glucan, β-glucan characterization, functional properties of β-glucan, FTIR analysis of β-glucan and how β-glucan can influence glucose and lipoprotein profile to provide health benefits.

Chapter 5 - Research on oat (*Avena sativa)* has been intensified in the last years as it has been reported to reduce serum cholesterol levels and attenuate postprandial blood glucose and insulin responses, which has been related to the presence of β-glucan. *A. sativa* is extensively planted as a forage crop in Northern Mexico, where rainfall has an erratic distribution resulting in smaller oat grains, which failed to meet the requirements of the market. In this regard, the extraction of a β-glucan-enriched oat gum from drought harvested *A. sativa* seeds could be an interesting alternative for this agricultural product to still represent value to producers. Mexican oat genotypes are important animal feed resources and have been studied on the basis of forage yield and nutritional value. Nevertheless, to our knowledge, studies on the extraction and characterization of oat gum from *A. sativa* Mexican genotypes harvested under different irrigation conditions have not been reported elsewhere. This chapter has been focused on the extraction and evaluation of the physic-chemical and gelling properties of β-glucan enriched oat gums from three oat genotypes (Karma, Cevamex and Cuauhtémoc) under two irrigation conditions: rainfed farming (RF) and irrigated crop (IC). Oat gums from RF Karma, Cevamex and Cuauhtémoc presented a β-glucan content of 62, 56 and 54 % (w/w), while lower values where found for IC genotypes (61, 50, 47 % w/w, respectively). The intrinsic viscosity $[\eta]$ and viscosimetric molecular weight (Mv) values of oat gums extracted from RF samples were higher than those of the corresponding IC genotypes. The oat gums formed physical gels (10% w/v in water) after heating at 75°C for 1 h and cooling at 25°C for 2 h. Large deformation mechanical tests (compression mode) revealed an increase in hardness of oat gums gels with increasing $[\eta]$ and Mv.

In: Oats: Cultivation, Uses and Health Effects ISBN 978-1-61324-277-3
Editor: D. L. Murphy, pp. 1-50

Chapter 1

OAT Β-GLUCAN, MOLECULAR CHARACTERISTICS AND HEALTH BENEFITS

***Jia Wu*[1] *and Bijun Xie*[2]**
[1]College of Biological Science and Technology, Fuzhou University, Fuzhou 350108, China
[2]College of Food Science and Technology, Huazhong Agricultural University, Wuhan 430070, China

ABSTRACT

Oat β-glucan is a linear polysaccharide composed of D-glucopyranosyl residues arranged as mixtures of consecutively β-(1→4)-linked glucose units in blocks that are separated by single β-(1→3) linkages. As a soluble dietary fiber, oat β-glucan has received extensive attention for its physiological effects on lowering cholesterol and postprandial blood glucose levels. The benefits are generally ascribed to the viscosity that formed in gastrointestinal tract by oat β-glucan. The viscosity depends on the solubility, molecular weight, and structure of oat β-glucan. So these molecular characteristics and the effect of processing on which are one of the focus of this review. The quantitative relationship between the hypoglycemic or hypocholesterolemic activities and the physicochemical characteristics of oat β-glucan in food matrix will be discussed. Since cereal β-glucan is an ingredient in functional food, many processes have been developed to produce β-glucan concentrates and isolates, which are widely applied in food industry.

Like mushroom-derived (1→3)-β-glucan, oat β-glucan displays immune activity, the recent discoveries on immune function will be given a brief introduction.

Abbreviations Used

AFM, atomic force microscopy;
AUC, area under the curve;
CE, Capillary electrophoresis;
CP-MAS NMR, cross-polarization magic-angle spinning nuclear magnetic resonance;
DP, degree of polymerization;
DP3, 3-O-β-cellobiosyl-D-glucose;
DP4, 3-O-β-cellotriosyl-D-glucose;
DSC, differential scanning calorimetry;
GC-MS, gas chromatography-mass spectrometry;
G', storage modulus;
G'', loss modulus;
HDL, high density lipoprotein;
HPAEC, high performance anion-exchange chromatography;
HPLC, high performance liquid chromatography;
HPSEC, high performance size exclusion chromatography;
LDL, low density lipoprotein;
LLS, laser light scattering;
MALDI-TOF MS, matrix assisted laser desorption ionisation time-of-flight mass spectrometry;
MS, mass spectrometry;
NMR, nuclear magnetic resonance;
PAD, pulsed amperometric detector;
PBGR, peak blood glucose rise;
RI, refractive index;
SCFA, short chain fatty acid.

1. Introduction

Mixed linkage (1→3)(1→4)-β-D-glucan (β-glucan) exists in oat, barley and other cereals. Oat β-glucan is a typical soluble dietary fiber. The primary locations of oat β-glcuan are aleurone cell walls and outer endosperm (Wood et al., 1983; Fulcher and Miller, 1993; Autio et al., 2001).

Traditionally oat β-glcuan was considered as an antinutritional factor in animal or poultry feed and it could reduce feed digestibility and animal or poultry body weight gain (Annison, 1991; Jeroch and Danicke, 1995; Bergh et al., 1999). But oat β-glucan is a functional nutrient instead of an antinutritional factor in the case of human health in rich countries. Clinical trials indicated that oat β-glcuan, as a soluble dietary fiber, is capable of lowering cholesterol and decreasing postprandial glucose levels in serum (Wood et al., 2000; Kerckhoffs et al., 2003; Karmally et al., 2005; Naumann et al., 2006). The US Food and Drug Administration (FDA) approved a health claim in 1997 that the use of oat-based foods can lower the risk of heart disease (FDA, 1997). These beneficial physiological functions are ususlly ascribed to the increased viscosity and adsorption of cholesterol by oat β-glcuan in human gastrointestinal tract (Guillon and Champ, 2000; Kerckhoffs et al., 2003; Wood, 2004; Queenan et al., 2007).

The viscosity depends on the physiological solubility, molecular weight, and structure of oat β-glcuan. So a lot of investigations on the molecular structure, rheology, and solubility in physiological conditions have been carried out to elucidate the relationship between oat β-glcuan characteristics and its physiological functions (Wood et al., 1991a; Roubroeks et al., 2001; Ren et al., 2003; Kerkhoffs et al., 2003; Lazaridou et al., 2004). The effect of oat β-glucan on immune system has also been revealed recently.

Ramakers et al. (2007) found that oat β-glucan could stimulate immune response in enterocytes. Specific benefit of soluble oat β-glucan on lung tumor metastases and macrophage antitumor cytotoxicity was reported (Murphy et al., 2004).

2. Enrichment and Extraction

Since the approval of the health claim by US FDA for oat β-glucan, the demand for oat fractions enriched in β-glucan has been on the rise. However, β-glucan exists in relative low concentrations in oat. Incorporation of oat β-glucan into food fomular at physiologically effective levels is a problem. Therefore, industry interest has been growing to produce oat β-glucan concentrates or isolates. Dry and wet processing have been developed to obtain oat β-glucan concentrates in large scale. Since molecular weight and solubility of oat β-glucan have great impact on the viscosity and the related physiological functions (Wood, 2004). Care must be taken to preserve the molecular characteristics of oat β-glucan. The dry processing can provide β-glucan concentrates with the content up to 30%, including both soluble and insoluble β-glucan. The concentrates are believed to offer the combined beneficial physiological effects of soluble and insoluble dietary fiber when incorporated into food systems. On the other hand, increased β-glucan solubility is observed in the extracts obtained by wet processing.

Doehlert et al. (1997) prepared β-glucan enriched oat bran by three dry milling processes. Roller milling and impact milling enriched β-glucan by 1.7 fold. If the oat were heated to 12% moisture, the yield of oat bran from roller milling could be improved nearly 2 fold. However pearling milling generated oat bran only slightly enriched in β-glucan. The result from pearling milling suggested that the distribution of β-glucan was fairly uniform throughout the groat used in the experiment. It was not as expected that β-glucan primarily located at outer layers of oat groat with low β-glucan content (Fulcher and Miller, 1993). The impact force in milling process crushes starch granules into fine particles, whereas the β-glucan enriched cell walls hold together as larger pieces, which can be separated by the sieving process. Heating appeares to strengthen the cell walls so that the yield of larger size bran fractions tend to increase. Further dry milling and sieving of oat bran produce fractions by size, the fraction with the largest particles ususlly had the hightest β-glucan concentration (Knuckles et al., 1992; Wu and Doehlert, 2002). Since the ground full-fat oat particles tend to clog the sieve and greatly decrease the yield (Knuckles et al., 1992; Wang et al., 2007), defatted rolled oat and oat bran are better materials for producing β-glucan concentrates (Åman et al., 2004). Air classification is an alternative technique to prevent the clogging of sieves by fine particles. In addition, air classification can easily be scaled up by using a commercially available large air classifier (Wu and Doehlert, 2002). Although dry process is a simple technique in industrial production, the produced concentrates usually have relative high protein and starch but low (< 30%) β-glucan content.

Wet processing is a technique to obtain concentrates with high β-glucan content. Aqueous solution is the most common solvent used for extracting β-glucan from cereals. Hot water, sodium carbonate, low concentration sodium hydroxide and dilute acid are suitable solvent for extraction. But concentrated acid, alkali, and high temperature should be avoid to prevent the degredation and hence the decrease of viscosity of β-glucan (Carr et al., 1990; Wood et al., 1991a; Bhatty, 1992; Bhatty, 1995). Aqueous alkali extraction, a typical wet processing, was developed by Wood et al. (1978). The authors obtained oat gum containing about 80% β-glucan. Sodium carbonate was used to adjust the pH of the slurry of oat fluor to pH 10. After extraction, the mixture was centrifuged to obtain the supertanant. Then the pH was adjusted to pH 4.5 to precipitate protein. The β-glucan remained in the supertanant after centrifuge was precipitated with 2-propanol. Then the extract was washed with 2-propanol to replace the water in the extract and air-dried with gentle warming. During their laboratory extraction, the effects of flour particle size, temperature, pH, and ionic strength of the extraction solution on the yields were investigated. On the basis of this laboratory trial, they also established a large scale extraction procedure (Wood et al., 1989). The resulted β-glucan exhibited significantly lower viscosity than that produced in laboratory scale. They found that the major viscosity loss occurred during centrifugation. It indicated that high shear force may break the molecular chain of β-glucan into shorter pieces and lead to the decrease of the viscosity. It is also possible that the residual enzyme activity might be responsible for the viscosity loss (Wood et al., 1989). Since endo β-1,3-glucanase, endo β-1,4-glucanase, and endo β-1,3-1,4-glucanase remained significant activity after alkali extraction (Wood et al., 1978, Beer et al., 1996; Beer et al., 1997a). Enzyme deactivation is necessary for producing high viscosity oat β-glucan. Hot aqueous ethanol refluxing is an effective method to inactive the enzymes. It seemed that defatting had little effect on the yield of β-glucan in producing β-glucan concentrates (Wood et al., 1978). But defatting was often used to obtain high purity β-glucan (Westerlund et al., 1993). Elevated temperature increased the yield of β-glucan (Wood et al., 1991b; Zhang et al., 1998), however gelatinized starch was coextracted with β-glucan and lowered the purity of the β-glucan extract. So thermostable α-amylase was usually added to hydrolyze starch at higher temperature. The residual protein and fat were eliminated by pancreatin (Lazaridou et al., 2003; Lazaridou et al., 2004).

Although wet processing yielded a high purity β-glucan product, the viscosity of the resulted β-glucan solution was relatively low. Since treatment at 80°C in 75% ethanol reduced the β-glucanase activity by only 12% (Beer et al., 1996), the β-glucan was hydrolyzed by those residual enzyme activity during aquous extraction. In order to solve this problem, an aqueous alcohol based enzymatic process was introduced recently by Vasanthan and Temelli (2008) to reduce the effect of β-glucanase during extraction process. This procedure separated starch and protein from unsoluble β-glucan in aqueous ethanol. The initial material was dispersed into 50% aqueous ethanol and screened to remove starch and protein (in filtrate). The β-glucan enriched fiber concentrate retained on the screen was redispersed in 50% ethanol again.

Protease and thermostable α-amylase were added sequentially followed by filtration to separate the hydrolysate. The β-glucan concentrate was retained on the screen. Cereal β-glucan molecules remained intact within the cell wall and were protected from the β-glucanase and shear fragmentation. Thus the β-glucan concentrates obtained by this technology showed superior rheological and physiological functions.

Alcohol is often used to precipitate oat β-glucan from aqueous solutions in the process of recovering. But the cost of recovering alcohol in industrial production is high. So alcohol free methods were developed. Ultrafiltration was an alternative of alcohol precipitation (Beer et al., 1996), and it consumed less time than dialysis. But the high shear force during ultrafiltration led to decreased viscosity and molecular weight of β-glucan. When β-glucan experienced freezing and thawing cycles, it formed cryogel. Which was a gelatinous or fibrous precipitate and could be separated by simple filtration (Morgan and Ofman, 1998). The technology has been patanted and the registered trademark of the produced β-glucan is "Glucagel".

3. Content Determination

With the increasing interest in oat and other cereal β-glucans as functional food components, there is a need for determining the content of β-glucan in grain and food. The most widely used method is an enzymatic procedure (McCleary and Codd, 2006). The β-glucans extracted from cereal and food were hydrolyzed by lichenase followed by β-glucosidase. The resulted glucose was determined by glucose oxidase/peroxidase and a chromogenic substrate. The method requires relatively simple instrument. But the cost for the reagent kits is high for high throughput analysis. In addition, high purity of the lichenase is critical for accurate determination.

Another method for determination of β-glucan measures fluorescence increase of Calcofluor/β-glucan complex. Calcofluor is a fluorescent dye capable of binding with cellulose and often used to localize cellulose and chitin (Haigler er al., 1980). Since oat β-glucan is a linear mixed linkage (1→3)(1→4)-β-glucan, which is similar with the cellulose molecular structure, it is reasonable that Calcofluor can bind with β-glucan. Wood and Fulcher (1978) found that Calcofluor could specificly bind to cereal β-glucans and cause precipitate in aqueous solution for the first time. The Calcofluor fluorescence method often involves the use of flow injection analysis (FIA). FIA method requires simpler procedure but more expensive FIA instrument.

The binding between Calcofluor and β-glucan is dependent on the ionic strength, the molecular weight of β-glucan, sodium chloride, and temperature. The molecular weight of β-glucan below a critical value lowered the fluorescence intensity of the complex. The fluorescence intensity increased with higher ionic strength (Kim and Inglett, 2006).

The factors influence the binding of Calcofluor and oat β-glucan were investigated by Wu et al. (2008). An alkaline buffer such as sodium carbonate is essential to solubilize Calcofluor and make a clear solution. Increasing the concentration of sodium carbonate buffer had two aspects of impact on the binding of Calcofluor to β-glucan, including enhanced binding ability and decreased potential binding sites.

Sodium chloride is usually added to the Calcofluor/β-glucan solution to increase the fluorescence intensity. Higher sodium chloride concentration (0.2 M) reduced the negative charge density on the surface of the β-glucan molecules and it is in favor of the binding of negatively charged Calcofluor on the β-glucan. The higher salt concentration also increased the dielectric constant of the solution and enhanced the binding of Calcofluor on β-glucan molecules. However, 1 M sodium chloride considerably suppressed the precipitation of β-glucan by Calcofluor (Wood and Fulcher, 1978). Although the precipitation of cereal β-glucan with Calcofluor is not appreciably affected by temperature from 5 °C to 25 °C (Wood and Fulcher, 1978).

The higher temperature was not favorable for the binding when the concentration of β-glucan solution is below the value that required for precipitate formation. The association of Calcofluor and β-glucan is driven by both enthalpy and entropy. The binding process involves hydrogen bonding, van der Waals forces, and hydrophobic interaction. Ethanol and maltose were mentioned by Izawa et al. (1996) to be factors that interfered with the combination of Calcofluor with β-glucan.

Recently a simple and cheap microplate assay method for determination of barley and oat β-glucan content was introduced (Schmitt and Wise, 2009). The results showed that the Calcofluor fluorescence method could be adapted to a fluorescent microplate reader.

It provided a simple means to measure β-glucan content in cereal grains and malt with inexpensive reagents and commonly available instruments. Some colorimetric methods using a dye binding procedure were also developed and may be alternative choices (Freijee, 2005).

4. MOLECULAR WEIGHT AND STRUCTURE

Although the primary component of cereal β-glucan is glucose residue, it is not easy to elucidate the molecular structure of the β-glucan. Some structural characteristics such as molecular weight, ratios of tri to tetra oligosaccharide, β-(1→4) to β-(1→3) linkages, and cellooligosaccharides content contribute to our understanding of the fine structure of cereal β-glucan.

4.1. Molecular Weight

The molecular weight of a polysaccharide is an important characteristic that determines the physicochemical properties such as viscosity. However, polysaccharides are polydisperse molecules. Therefore the molecular weights of polysaccharides are averages and the distribution of molecular weight plays an equally important role. The most frequently used molecular weight averages are the weight-average molecular weight M_w and the number-average molecular weight M_n. Methods of analysis include various chromatographic techniques which require calibration with standards and absolute techniques like light scattering in which calibration is not required and no assumptions are made concerning the conformation of the molecule (Harding et al., 1991). High performance size exclusion chromatography (HPSEC) equipped with laser light scattering (LLS), refractive index (RI), viscosity detector or post-column binding with Calcofluor and fluorescent detection are often used to determine the molar weight of cereal β-glucans (Beer et al., 1997a; Rimsten et al., 2003; Åman et al., 2004). The molecular weight of linear cereal β-glucans can be determined according to the Mark-Houwink equation $[\eta] = KM^a$. where $[\eta]$ is the intrinsic viscosity, K and a are constants, the values of which depend on the nature of the polymer and solvent as well as on temperature, M is usually one of the relative molecular mass averages. The number-average molecular weight of oat β-glucan can also be determined by osmotic pressure measurements (Vårum et al., 1991).

The reported values of the molecular weight of different oat β-glucans vary between 6.5×10^4 and 3.1×10^6 (Doublier and Wood, 1995; Lazaridou et al., 2003; Lazaridou and Biliaderis, 2007a). The comparison of these values is difficult due to the strong dependence of the molecular weight on the methods of extraction and analysis. Extracting cereal material with increased solution temperature obtains β-glucan with higher molecular weight. For instance, The molecular weights of β-glucan extracted from oat grains at 40, 65, and 100 °C were 1.18×10^5 – 1.024×10^6 Da, 9.85×10^5 – 1.999×10^6 Da, and 2.3×10^6 Da respectively (Zhang et al., 1998). If the analysis system is not equipped with a laser light scattering detector, molecular weight standards have to be used to calibrate the HPSEC column. Since the retention time of the polysaccharide is dependent on hydrodynamic volume rather than molecular weight, the use of pullulan or dextran as standards can lead to significant overestimation of molecular weight of polysaccharides with a more rigid backbone or extended chain such as cereal β-glucans (Gómez et al., 1997a; Wang et al., 2003). Generally speaking, the molecular weight of oat (6.5×10^4 – 3.1×10^6) is higher than barley (3.1×10^4 – 2.7×10^6) (Morgan et al., 1999; Wood et al., 1994a; Lazaridou and Biliaderis, 2007a) and rye (2.1×10^4 – 1.1×10^6) (Roubroeks et al., 2000a; Wood et al., 1991a; Lazaridou and Biliaderis, 2007a).

Oat and barley β-glucans can be fractionated by stepwise ammonium sulphate precipitation according to the molecular weight of the fractions. The two kinds of β-glucans with broad molecular weight distribution were separated into seven fractions. The analysis indicated no difference in oligosaccharide pattern between the fractions and the initial β-glucans (Wang et al., 2003). However, the fractions of oat β-glucan prepared with the similar ammonium sulphate precipitation method showed different molecular structure. The ratio of tri to tetra oligosaccharide in the fractions lowered with decreased molecular size (Izydorczyk, et al., 1998a). The differecne between the above two separation characteristics was probably due to the association of β-glucan molecules in the different procedure. The solution was allowed to stand for only 1 h at 25 °C after desired concentration of ammonium sulphate was achieved in the procedure of Wang et al. (2003). While the solution was left overnight at 5 °C before centrifugation in the procedure of Izydorczyk, et al. (1998a). Since β-glucan with high content of cellulose like regions and continuous cellotriosyl units linked by β-(1→3) linkages tend to form aggregates with decreased solubility in the solution. The low temperature and prolonged time generally favors molecular association. It is reasonable that the fractions from Izydorczyk, et al. (1998a) have different molecular structure.

4.2. Structure

The water soluble β-glucan from oat consists of about 70% (1→4)-β-glucosyl residues and 30% (1→3)-β-glucosyl residues. The (1→3)-β-glucosyl residues are always flanked on both their reducing end and nonreducing end sides by (1→4)-β-glucosyl residues. Two or more adjacent (1→3)-β-glucosyl residues are not found. The β-(1→4)-linkages of most cereal β-glucans are arranged predominantly in groups of two or three adjacent linkages. In oat β-glucan, this arrangement accounts for about 90% of the polysaccharide (Woodward et al., 1983). The remaining 10% consists of longer blocks of from 4 up to 17 adjacent (1→4)-β-glucosyl residues (Wood et al., 1994a; Izydorczyk et al., 1998b). In other words, oat β-glucan can be viewed as β-(1→3)-linked cellotriosyl, cellotetraosyl, and cellulosic residues (Figure 1). The distribution of the two structural features 3-O-β-cellobiosyl-D-glucose (DP3) and 3-O-β-cellotriosyl-D-glucose (DP4) has been determined to be random by treating it as a Markov chain (Staudte et al., 1983). The later investigation indicated that the compositions of hydrolysate from a cellobiohydrolase II are well reproduced by a second order Markov chain (Henriksson et al., 1995). The ratio of DP3 to DP4 increases with oat (1.5 – 2.3) < barley (1.8 – 3.5) < wheat (3.0 – 4.5) (Lazaridou and Biliaderis, 2007a).

The most convenient method to obtian these oligosaccharides in analysis of the structure of β-glucan is enzymatic hydrolysis . Lichenase and cellulase have been used in the hydrolysis of cereal β-glucans (Wood et al., 1989; Wood et al., 1991a; Henriksson et al., 1995; Johanssona et al., 2000b; Roubroeks et al., 2000b; Tosh et al., 2004a; Ajithkumar et al., 2006). Lichenase is a (1→3)(1→4)-β-D-glucan 4-glucanohydrolase (EC 3.2.1.73) which specifically cleaves β-(1→4)-glucosidic linkages of a 3-substituted unit, thus yielding oligosaccharides containing a single (1→3)-linkage adjacent to the reducing end (Figure 1). Complete enzymatic hydrolysis of oat β-glucan yields the oligosaccharides DP3 and DP4. These two components account for 90% of the polysaccharide. Cellulase is an endo-(1→4)-β-D-glucan-4-glucanohydrolase (EC 3.2.1.4) which hydrolyzes the internal (1→4)-linkages next to 4-substituted glucose residues.

To indentify the structure and content of the oligosaccharides released by hydrolyzing enzymes, chromatography, mass spectrometry (MS), and nuclear magnetic resonance (NMR) have been used in analysis. Wood et al. (1991a) developed a HPLC technique to quantify the oligosaccharide hydrolized by lichenase from cereal β-glucans. The oligosaccharides were first separated on a high performance liquid chromatography (HPLC) column and then detected by orcinol-sulfuric acid reaction. However HPLC can not adequately resolve the oligosaccharides with degree of polymerization (DP) > 4.

In order to overcome this problem, high performance anion-exchange chromatography with pulsed amperometric detector (HPAEC-PAD) was applied to analyze cellulose-like oligosaccharides released by lichenase from β-glucan (Wood et al., 1994a).

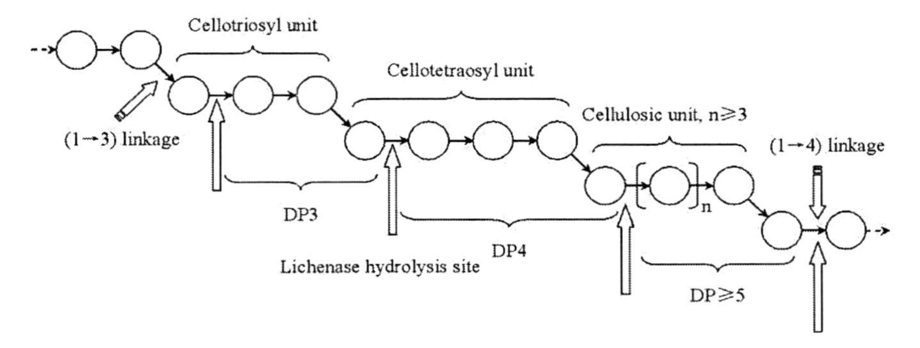

Figure 1. Schematic structure of oat β-glucan molecule. The circles represent glucose residues. The vertical arrows indicate the hydrolysis sites of lichenase on oat β-glucan molecule. DP3: 3-O-β-cellobiosyl-D-glucose; DP4: 3-O-β-cellotriosyl-D-glucose; DP≥5: cellulosic oligosaccharides containing more than three consecutive 4-O-linked glucose residues.

Then, it became the most widely used method in oligosaccharide components analysis for β-glucan hydrolysates. Capillary electrophoresis (CE) provides a sensitive and selective analysis method for the structural components of β-glucan (Johansson et al., 2000; Colleoni-Sirghiea et al., 2003). More rapid and sensitive method such as matrix assisted laser desorption ionisation time-of-flight mass spectrometry (MALDI-TOF MS) have been used to quantify oligosaccharides released from water-soluble barley β-glucan by lichenase hydrolysis (Jiang and Vasanthan, 2000). MALDI-TOF MS takes about 20 min for determining peak heights for 10 probe positions rather than 1 h for each analysis by HPAEC-PAD. The sample concentration used by MALDI-TOF MS was 1/12 of that used by HPLC or HPAEC-PAD. MALDI-TOF MS was also applied in analyzing the oligosaccharides of *Equisetum arvense* hydrolized by lichenase (Sørensen et al., 2008).

Since it is difficult to obtain pure oligosaccharides released by lichenase, the quantification of oligosaccharides by HPAEC-PAD was limited to the knowledge of weight response factors (Wood et al., 1994a). In general, the sensitivity of PAD decreased rapidly from DP2 to DP6, while, for higher molecular weight oligosaccharides (DP7-17), the decrease in the sensitivity of PAD was minimal (Timmermans et al., 1994). Two major lichenase-hydrolyzed products were generated from the barley β-glucan, and DP3 and DP4 were separated and highly purified by recycling preparative HPLC technology (Yoo et al., 2007). The authors found that the amount of DP3 and DP4 accounted for only 71.4% to 73.3% of water-extractable barley β-glucan fraction, but the ratio of DP3/DP4 was in accordance with other reported studies. The lower total amount of DP3 and DP4 may be ascribed to the higher total β-glucan content determined by the authors.

Gas chromatography-mass spectrometry (GC-MS) is useful in determination of the oligosaccharide structure in β-glucan molecules. The methylated oligosaccharides were hydrolysed and converted to alditol acetates, then analysed by GC-MS. The results showed that losses occured at the reducing end during methylation (Wood et al, 1991a). The completely methylated oat β-glucans were also analyzed by GC-MS and the ratios of β-(1→4)-linkage to β-(1→3)-linkage were determined (Izydorczyk et al., 1998a, 1998b).

The structure of β-glucan can be investigated by NMR spectroscopy. Dais and Perlin (1982) proposed primary structure for β-glucans (Figure 2). The authors pointed out that β-glucans were composed of four different glucose residues, corresponding to A, B, C, and D in Figure 2. As the signals for C-4 and C-3 engaged in the interresidue linkages are well separated, their integrals were used to estimate the ratios of (1→4)- to (1→3)-linkages. The value was 2.2 for lichenan and 2.6 for oat and barley β-glucan. The structure features were confirmed by later research results (Lazaridou et al., 2004; Cui et al., 2000; Johansson et al., 2000; Roubroeks et al., 2000a; Wood et al., 1994a). Using DMSO-d_6 instead of D_2O at elevated temperatures was known to give better resolution (Dais and Perlin, 1982). Cui et al. (2000) reported first characterization of a wheat β-glucan by NMR. Direct and long-range homo-($^{1}H/^{1}H$) and hetero-($^{13}C/^{1}H$) nuclear shift correlations were used to make complete assignments of both the ^{13}C and ^{1}H spectra as well as to confirm sequences and linkage sites.

In ^{13}C NMR spectra, the C-1 region is characterized by the presence of three resonances, which could be used to confirm the purity or homogeneity of β-glucan isolates (Irakli et al., 2004; Lazaridou et al., 2003). C-4 signals of A, C, and D rings are far from the C-4 signal of B ring (Wood et al., 1991a), so it is possible to estimate the ratio of (1→4)- to (1→3)-linkage by integral of the two groups of C-4 signals.

$$\begin{array}{cccccccccc} A & B & C & A & B & C & D & A & B & C \end{array}$$
$$\rightarrow {}^{4}G_1 \rightarrow [{}^{3}G_1 \rightarrow {}^{4}G_1 \rightarrow {}^{4}G_1 \rightarrow]_m [{}^{3}G_1 \rightarrow {}^{4}G_1 \rightarrow {}^{4}G_1 \rightarrow {}^{4}G_1 \rightarrow]_n {}^{3}G_1 \rightarrow {}^{4}G_1 \rightarrow$$

Figure 2. Generalized β-glucan structure.

Wood et al. (1994a) distinguished three distinct resonances from C-4 of A, C, and D ring. The two intense resonances arose from β-(1→4)-linked glucose residues (A and C ring) flanked on either the reducing or nonreducing end by β-(1→3)-linked glucose residue. The smaller resonance at higher field was from the more cellulose-like environment (D ring) in β-glucan. So the smaller resonance was a rough measure of the proportion of sequences with more than three consecutive β-(1→4)-linkages. The ^{13}C NMR spectra showed single resonances for the C-3 and C-4 of B ring, indicating a single environment for this residue in the polymeric structure and the absence of two consecutive (1→3)-linkages in the structure of β-glucans (Wood et al., 1991a; Lazaridou et al., 2003; Skendi et al., 2003). The double signals of C-6 corresponds to the C-6 of B residues and to the C-6 of A, C, and D residues. The relative intensities of the two resonances can thus be used to evaluate the ratio of (1→4)/(1→3) linkages in β-glucan. The value for oat β-glucan is between 2.34 – 2.75 (Dais and Perlin, 1982; Lazaridou et al., 2003; Skendi et al., 2003).

In the ^{1}H NMR spectra of β-glucan and its hydrolysates, the resonances beyond the crowded region from about 3.0 to 4.0 ppm can be used as structural reporter signals. These include the resonances for all anomeric protons, which have chemical shifts between 4.3 and 5.3 ppm. Quantitative analysis can also be carried out from the signals in this region by integration and their assignment is known previously (Dais and Perlin, 1982; Henriksson et al., 1995; Colleoni-Sirghiea et al., 2003). The anomeric proton resonances can be divided into three groups: (1) protons at reducing end with either α- or β-anomers; (2) β-(1→3)-linkage protons at non-reducing end or inside the β-glucan; and (3) β-(l→4)-linkage protons at non-reducing end or inside the β-glucan. The H-1 signals in ^{1}H NMR spectra can provide useful information in identifying the structure of oligosaccharides released by certain enzymes. When barley β-glucan was hydrolyzed by an exo-cellulase, the major hydrolysates were separated and identified to be cellobiose, 3-O-glucosyl-D-cellobiose, 3-O-glucosyl-D-cellotriose, and 3-O-cellobiosyl-D-cellobiose. The average oligosaccharide length was 3.1 after 72 h of hydrolysis, calculated from the amount of reducing ends as determined by ^{1}H NMR. The amount of β-(1→4)-linkages decreased from 71% to 40% while the amount of β-(1→3)-linkages kept constant at 29% (Henriksson et al., 1995). Roubroeks et al. (2000b, 2001) used both lichenase and endo-cellulase to hydrolyze oat β-glucan.

The hydrolysate released by lichenase after 70 h incubation included laminaribiose, 3-O-cellobiosyl-D-glucose, 3-O-cellotriosyl-D-glucose, and polymeric material. The laminaribiose maybe derived from other enzyme activity such as cellulase.

The major hydrolysis products by endo-cellulase were glucose, cellobiose, 4-O-laminaribiosyl-D-glucose, 4-O-laminaribiosyl-D-cellobiose, and 3-O-cellobiosyl-D-cellobiose as reported before (Bock et al., 1991; Henriksson et al., 1995; Shiroishi et al., 1997). However, the findings suggested that the 3-O-cellobiosyl-cellobiose may be degraded by endo-(1→3)-*β*-glucanase and produce cellobiose. And the high glucose content was partially ascribed to contaminent enzyme activity (Roubroeks et al., 2001). The ratio of (1→4) to (1→3)-linkages could be obtained from ^{1}H NMR, the value was 2.31 for *β*-glucan from rye bran and 2.4 for *β*-glucan from oat (Roubroeks et al., 2000a; Colleoni-Sirghiea et al., 2003).

Solid state ^{13}C CP-MAS NMR technique has been used to obtian direct information on the water insoluble fractions of oat and barley *β*-glucan (Virkkia et al., 2005; Johansson et al., 2004).

The results showed that resonances between 10 to 40 ppm, which were aliphatic origin, were derived from fat and protein residues (Pizzoferrato et al., 2000). Signals at 173 ppm were of carbonyl carbons from cell wall proteins (Davies et al., 2002). The resonance at 103 – 105 ppm was C-1 and signal at about 60 ppm was C-6. The other carbons C-2, C-3, and C-5 appeared at 74–75 ppm. The broad shoulder at about 80 ppm is from C-4.

5. Solution and Gel Properties

Since viscosity of cereal *β*-glucan is responsible for physiological benefits and thickening effects. It is of great importance to investigate the rheology properties of aqueous solution and gel containing oat and other cereal *β*-glucans. The flow or gel behavior of cereal *β*-glucan depends on shear rate, concentration, storage time, temperature, molecular weight, and fine structure.

5.1. Viscoelasticity

Viscosity describes the resistance to flow of a fluid which is deformed by shear stress. Viscosity-shear rate curves are often plotted to investigate the factors that have impact on the viscosity of cereal *β*-glucan solutions. Generally, cereal *β*-glucan solutions show Newtonian flow behavior at low shear rate and non-Newtonian flow behavior (shear thinning) at high shear rate above a certain concentration (Ren et al., 2003). At low shear rates, those molecular entanglements disrupted by the imposed deformation are replaced by new interactions between molecules, leading to no change in the density of entanglements, and hence, no reduction in viscosity (Morris et al., 1981). The flow behavior corresponds to the plateau region of the viscosity-shear rate curve. While at high shear rate, the speed of entanglements disruption exceeds that of new entanglements formation. Thus the structure in the solution decreases, leading to lower viscosity. It is reasonable that the viscosity of cereal β-glucan solution increases with the higher β-glucan concentration (Böhm and Kulicke, 1999a; Ren et al., 2003; Papageorgiou et al., 2005). The fresh oat β-glucan solution, especially with low molecular weight, exhibits typical random coil flow behavior, while stored for a period of time, it shows shear thinning behavior at low shear rate. The significant shear thinning behavior without a Newtonian plateau means the associations of ordered β-glucan molecules become enough to form intermolecular networks in solution (Miyoshi and Nishinari, 1999). There are two association mechanisms have been raised. One supposes that cellulose-like sequences tend to form interchain aggregation by hydrogen bonds (Woodward et al., 1983; Doublier and Wood, 1995). The other assumes that sequential cellotriosyl units are responsible for association of β-glucan molecules (Böhm and Kulicke, 1999b). X-ray fiber diffraction of barley β-glucan and lichenan indicated that three consecutive cellotriosyl units were in helical conformation and may form a stable complex between two β-glucan chains (Tvaroska et al., 1983). Study on structural and rheological characteristics suggested that β-(1→3)-linked cellotriosyl units in β-glucan molecules were probably the segments which formed the junction zones rather than cellulose-like segments (Tosh et al., 2004a, b). The time dependence of viscosity was also displayed in thixotropic loop experiments (Lazaridou et al., 2003). Higher temperature (40 – 50 °C) made the flow behavior more similar with Newtonian fluid. The viscosity-shear rate curve became more flat than that at lower temperature, in other words, it was a more Newtonian-like behavior. The effect of temperature on the flow behavior could be fitted with Arrhenius equatioin and the activation energy could be obtained (Papageorgiou et al., 2005). As expected, there is an increase in viscosity and shear thinning of oat β-glucan solution with increasing molecular weight (Lazaridou et al., 2003; Skendi et al., 2003).

The storage modulus (G') describes the stored energy, representing the elastic portion; and loss modulus (G'') describes the energy dissipated as heat, representing the viscous portion of viscoelastic fluid or gel. In typical cereal β-glucan solutions with moderate concentration and oscillation frequency, loss modulus is higher than storage modulus. As the oscillation frequency increases, the storage modulus increases faster than the loss modulus and above a certain frequency the storage modulus exceeds the loss modulus, and the solution property changes from a liquid-like to a solid-like behavior (Lazaridou et al., 2003; Ren et al., 2003). Storage and loss modulus usually are higher in concentrated solution than in dilute solution. Oat β-glucan solutions above a certain concentration can form gel when cured under room temperature. The aqueous solutions ($G' < G''$) gradually turn into gel ($G' > G''$) and the time when G' crosses G'' is usually considered as gelation time G_t. The maximum slope of log $G'(t)$ was defined as gelation rate $I_E = (d\log G'/dt)_{max}$ (Böhm and Kulicke, 1999b). The low molecular weight β-glucan solution has shorter gelation time and higher gelation rate than its high molecular weight counterpart due to higher mobility for smaller β-glucan molecules (Doublier and Wood, 1995; Lazaridou et al., 2004; Vaikousi et al., 2004). It is worthy of note that the storage modulus of β-glucan gel increases with higher DP3 content rather than cellulose-like segments. The average G' is linearly correlated with percentage for DP3 (Tosh et al., 2004b). The compression tests clearly showed an increase of gel strength as well as a decrease of gel brittleness for β-glucan with higher molecular weight, ratio of DP3 to DP4, and concentration. (Lazaridou et al., 2004; Vaikousi et al., 2004). Cereal β-glucan gel will gradually melt when heated. The G' decreases sharply near melting point at which $G' = G''$ (Papageorgiou et al., 2005). An alternative technique to determine the melting point is differential scanning calorimetry (DSC). The endothermic peaks in DSC curve become broader and shift to higher temperatures as the cellotriosyl content increases. Analysis indicated that the peak and onset temperature were positively correlated with DP3 content (Tosh et al., 2004b). The apparent melting enthalpy increased with decreasing molecular size. Whereas the melting temperature showed the reverse trend (Lazaridou et al., 2003). The opposite phenomina indicates that there are better organization, probably caused by long chain segments in the gel containing high molecular β-glucan. And there are more junction zones in gel containing low molecular weight β-glucan. Results from ^{13}C CP/MAS NMR spectroscopy showed that a kind of barley β-glucan contained regions with two distinct conformations. In some of the regions the β-glucan chains associated to form a unique conformation, while in the other regions the β-glucan chains were in an amorphous conformation (Morgan et al., 1999).

It is interesting that cereal β-glucan aqueous solution can form cryogel after freezing-thawing cycles (Morgan and Ofman, 1998; Lazaridou and Biliaderis, 2004; Vaikousi and Biliaderis, 2005). The major reason for the cryogelation is solute concentrating in those unfrozen parts of the apparently frozen solid system that remained below 0 °C. When any solution freezes at moderate temperature below 0 °C, the crystals of a pure solvent are initially formed. This causes an increase in the solute concentration in the still unfrozen liquid regions of the system, thus resulting in a strengthening of polymer-polymer interactions and the formation of condensed polysaccharide matter. The cryogelation process can be divided into two stages. The formation of initial cryoprecipitates is induced by freezing the system only, which could be considered as the first stage of cryostructuring. The thawing step could be considered as the second stage of cryostructuring. This second stage is responsible for the high efficiency of the process. Slow thawing process could promote the formation of cryostructures (Lozinsky et al., 2002). The cryogelation phenomena of milk powder, barley β-glucan, and polyol mixture dispersed in aqueous solution were investigated by Lazaridou and Biliaderis (2007b). The addition and increasing concentration of polyols (fructose, glucose, sucrose, and xylose) to β-glucan solutions retarded the cryogelation of barley β-glucan solution and resulted in weaker cryogels. Whereas sorbitol seemed to promote the formation of cryostructure of β-glucan (Lazaridou et al., 2008).

5.2. Aggregation in Dilute Solution

Except gel formation ability in concentrated aqueous solution, β-glucan tends to form aggregates in dilute solution. Vårum et al. (1992) found that the aggregares was labile and could be completely dissociated into monomers on dilution approaching zero concentration, which was not inconsistent with the report by Li et al. (2011), who considered that the aggregation was a fast process and the equilibrium between association and dissociation reached quickly. This was supported by the results that heat treatment, filtration, ultrasonication, and the use of urea solution could not eliminate aggregates completely (Li et al., 2006a). Light scattering and rheology are the two useful techniques in aggregation behavior investigation. It is necessary to obtain β-glucan monomers in quantitative study of the aggregates. Several solvents and methods were effective in producing monomers. Cuoxam, a solvent for dissolving cellolose, was found to be capable of dispersing cereal β-glucan into monomers (Grimm et al., 1995). The viscosities of oat and barley β-glucan at concentrations of 15 mg/ml in cuoxan were much lower than in pure water at low shear rates but approximately the same at high shear rates. In cuoxan, no shear thinning was observed (Johansson et al., 2008). This shows that β-glucan is molecularly dispersed in cuoxam and no association or entanglement occurs at this concentration. In 0.5 M NaOH solution, no wheat β-glucan aggregates were detected by dynamic light scattering measurement (Li et al., 2006a). It was found from light scattering data that the worm-like cylinder model could explain the chain stiffness better than the random flight model for wheat β-glucan in 0.5 M NaOH solution. Which suggested an extended random coil conformation of β-glucan molecules in 0.5 M NaOH solution. (Li et al., 2006b). Barley β-glucan chain was also modelled by a partially stiff worm-like cylinder by viscosity measurements of barley β-glucan solutions (Gómez et al., 1997b). Single oat β-glucan molecules were visualized by atomic force microscopy (AFM) using surfactant for dispersing the molecules (Wu et al., 2006, 2009). Derivation is effective in elimination of β-glucan aggregates. The carbanilation of cereal β-glucans can prevent intermolecular hydrogen bonding by blocking hydroxyl groups with phenyl carbamate groups (Li et al., 2007). Microwave heating can disperse cereal β-glucan completely in aqueous solution. When β-glucan dispersed in water was heated by microwave at 100 to 121 °C for 4 to 10 minutes, the polysaccharide solubilized completely withour degradation (Wang et al., 2002).

Aggregation states of β-glucan depend on the solvent. Cuoxam is a good solvent and can disperse β-glucan molecules into monomers, while water is a poor solvent and causes severe aggregation of β-glucan molecules. Higher β-glucan concentration leads to larger aggregates in aqueous solution (Grimm et al., 1995; Wu et al., 2006). The association level is enhanced by an increase in temperature. Addition of NaCl would weaken the hydrogen bonds between β-glucan molecules and decrease the association level (Gómez et al., 1997c). Molecular weight and structure are important factors that can impact the aggregation of β-glucan. Since aggregation in dilute solution is diffusion limited. Low molecular weight β-glucan molecules increase the aggregation level due to higher diffusion rate. While β-glucan with higher DP3 to DP4 ratio is more rigid and has a lower diffusion rate, leading to a decreased aggregation level (Li et al., 2011). The aggregates in oat β-glucan aqueous solutions with different concentration and storage time were investigated by AFM (Wu et al., 2006). The results showed that larger spherical aggregates formed in solutions with higher β-glucan concentration up to 50 μg/mL. The microfibrous structures over 3 μm long existed in 100 μg/mL β-glucan solution. Storage of low concentration β-glucan for 9 days also led to the formation of larger aggregates. The authors also visualized large fibrous aggregates by confocal scanning laser microscopy.

Some aggregation models has been proposed. Vårum et al. (1992) tried to fit their data from oat β-glucan to a spherical micelle model. While the findings from light scattering and viscometry indicated fringed micelle formation in barley β-glucan solution (Grimm et al., 1995). The fringed micelle aggregation was visualized by AFM (Wu et al., 2006). The growth of aggregates leads to branched structure. Since the growing branches prevent the molecules from entering the inner part, the more opened structure of aggregates is formed. Based on the concentration dependence of the average apparent diameters of cereal β-glucans, Li et al. (2011) chaimed that the cluster-cluster aggregation was dominant in the solution.

6. Changes in Germination and Processed Food

6.1. Changes in Germination

Germination can be used to enhance the flavor and texture of cereals. However, germination usually causes breakdown of β-glucans, leading to a decreased physiological function. So it is necessary to study the germination process and find a compromise between sensory property and physiological effect. The effect of malting on barley β-glucan was investigated on three factors, steeping temperature (15 and 48 °C), moisture content (38%, 42%), and germinatioin temperature (15 and 18 °C) by Rimsten et al. (2002). The authors found that the activity of β-glucanase could be reduced by high steeping temperature (48 °C). Steeping at 48 °C also cost shorter time to reach desired moisture content and decreased the hydrolyzing time for the enzyme. The other two factors influenced the outcome to a small extent. The results from extracting β-glucan at various temperatures indicated that the β-glucan extractable at 100 °C but not at 38 °C was the main substrate for β-glucanase. Prolonged germination time could decrease the content of β-glucan. Peterson (1998) found that nearly all the β-glucan was degraded after germination at 16 °C for 6 days. The analysis for germinaed oat by Hübner et al. (2010) also confirmed that longer germination periods considerably decreased the content of β-glucans. While germination temperature between 10 and 20 °C showed little effect on β-glucans. The growth of microbes is an important factor that lowers the quality of the germinated cereal grains. A hulled oat and a naked oat were germinated at 5, 15 and 25 °C. The results showed that elevated temperature led to an increase in the amount of bacteria. Lower germination temperature should be used to avoid the excessive growth of microbes and the activity of β-glucan hydrolyzing enzymes. A short germination time and low temperature (72 h, 15 °C) followed by oven drying could produce germinated oat with 55% to 60% of initial β-glucan content, the molecular weight of which was slightly lowered (Wilhelmson et al., 2001).

6.2. Changes in Processed Food

Although processing may increase the solubility of β-glucan in food, the molecular weight of β-glucan in processed food is usually lower than that in unprocessed whole grains. The physical state of β-glucan in the raw material, food matrix, β-glucanase activity, processing, and storage conditions affect the physiological solubility (availability) and molecular weight of β-glucan in food system. An *in vitro* digestion system is often used to simulate human digestion. The extracted cereal β-glucan by *in vitro* digestion system reflects the solubilization and viscosity formation of β-glucan in human gut.

Heating oat bran slurry may increase the amount of physiological soluble β-glucan, for instance, from 29% to 84% (Jaskari et al., 1995). The authors also found that the molecular weight of β-glucan remained unchanged during the hydrothermal and α-amylase treatments. While the β-glucan could be degraded by β-glucanase activity in phytase. In some situations, hydrothermal treatment can not increase the amount of soluble β-glucan. The results form Izydorczyk et al. (2000) showed that autoclaving and steaming treatments had no effect on the extractability of β-glucans from barley, but prevented enzymatic hydrolysis of β-glucans. It should be noted that the soluble β-glucan content was determined by aqueous extraction at 25 °C, which may be different from a physiological condition. The addition of protease and esterase during extraction as well as sonication treatment increased extractability of β-glucans from barley (Izydorczyk et al., 2000). Extractability of β-glcuan by *in vitro* digestion was higher for muffin than the oat bran, whereas the molecular weight decreased in muffin (Beer et al., 1997b). Analysis of β-glucan from porridge and bread showed that cooking (for porridge) increased the amount of soluble β-glucan but baking (for bread) decreased it (Johansson et al., 2007). The decrease of soluble β-glucan content in bread may caused by enzyme activity in the wheat flour or in the added yeast towards β-glucan. Åman et al. (2004) found decreased molecular weight of β-glucan in bread, which was also ascribed to enzymatic hydrolysis in fermentation step. Hot air drying decreased the amount of soluble β-glucan in bread and fermentate, but for porridge the amount of soluble β-glucan increased in drying (Johansson et al.,

2007). Although solubility and molecular weight of β-glucan may change in food production. Processing did not result in differences in the DP3 to DP4 ratio of soluble β-glucans (Johansson et al., 2007). Deactivation of enzyme activity in oat kernels may be a useful method to retain soluble β-glucan. Deactivation treatments such as steaming, autoclaving, hot air roasting and infrared roasting did not cause significant loss of oat β-glucan (Hu et al., 2010).

Kivelä et al. (2010) found that irreversible fall in viscosity and a loss in shear thinning behaviour occurred in oat β-glucan solutions as a result of homogenisation treatments. High shear force may be responsible for the loss of viscosity (Wood et al., 1989). Increasing the temperature and standard mechanical energy, a measure of the energy input during the extrusion process, caused depolymerization of β-glucan during extrusion. By this method, Tosh et al. (2010) prepared a series of extruded oat bran cereals in which the molecular weight distributed from 1.9×10^6 to 2.5×10^5, the solubility increased with lowering molecular weight of the sample. Except enzyme and acid hydrolysis (Vaikousi and Biliaderis, 2005) of cereal β-glucans in food processing, oxidation is also an important factor that can cause degradation of β-glucans. Kivelä et al. (2009a, b) found ascorbic acid induced oxidative cleavage in modelled beverage conditions. The inhibitors prevented and the metals effectively catalysed the viscosity decrease of the beverage. Which suggests oxidative degradation is a potential threat for the stability of β-glucan in certain fiber enriched products.

The authors also pointed out that citric and malic acid affected the molecular weight and vicosity of β-glucan moderately after a heat treatement such as a pasteurizing process. Cereals provide suitable growth media for lactic acid bacteria.

Sometimes the amount of soluble β-glucan increased after fermentation (Degutyte-Fomins et al., 2002), but other reports showed a decrease in the amount of β-glucan (Mårtensson et al., 2002; Lambo et al., 2005). It is interesting that the molecular weights of β-glcuan were not significantly affected by fermentation (Lambo et al., 2005). The above results reflected a dynamic change in the process of fermentation.

During frozen storage, extractable β-glucan may decrease by >50%, but no change in molecular weight of extracted β-glucan was detected (Beer et al., 1997b). Freeze-thaw cycling process for beverage and food enriched with β-glucan may cause cryogel, which could lead to the formation of precipitates in beverage and the changes of viscosity or sensory property of food during processing and storage. (Vaikousi and Biliaderis, 2005).

7. Health Effect

As a kind of soluble dietary fiber, cereal β-glucans have received increasing attention for the health benefits in lowering plasma cholesterol and postprandial blood glucose levels. Numerous clinical studies have verified the physiological benefits of β-glucans and led to the approval of health claims that oat and barley β-glucans can reduce the risk of coronary heart disease (FDA, 1997, 2005). The major reason for blood glucose and cholesterol lowering effect is ascribed to viscosity of β-glucan that formed in mixed digesta after ingestion. The viscosity of β-glucan is related to the molecular weight and concentration in physiological state. The other hypothesis include cholesterol and bile acid binding (Kahlon and Woodruff, 2003), and short chain fatty acid (SCFA) by fermentation (Queenan et al., 2007). The physiological effects of cereal β-glucan could be varied by different food matrix and processing (Regand et al., 2009; Kerckhoffs et al., 2003). Cereal β-glucan also showed immune modulation function in recent research (Volman et al., 2008). In the large bowel, β-glucans have the potential to become sources of carbon and energy for probiotics such as lactobacilli and thus to boost bacterial numbers or metabolism (Snart et al., 2006).

7.1. Effect on Blood Glucose

The most important factor that influence the function in lowering blood glucose by cereal β-glucan is viscosity. Cereal β-glucans increase the viscosity of the intestinal contents, causing reduced postprandial glucose and insulin levels. Then the lower ambient insulin levels improve the cellular insulin sensitivity, resulting in improved glucose metabolism. The hypoglycemic effects of β-glucans have been reported by a large amount of literatures (Casiraghi et al., 2006; Hlebowicz et al., 2008). Vachon et al. (1988) are probably the first authors that reported the relation between oat β-glucan and postprandial glucose and insulin levels. The authors also speculated that the effect of soluble fibers was closely related to the viscosity of fiber solutions in the presence of the diet ingredients.

Varying dose, molecular weight, and viscosity of β-glucan may cause different physiological effects. Hooda et al. (2010) used a porto-arterial catheterization model to study the effect of oat β-glucan on glucose and insulin levels. The results showed that when pigs were fed three diets containing 0%, 3% and 6% oat β-glucan, only the diets with 6% β-glucan decreased net glucose flux and peak apparent insulin production significantly. Tappy et al. (1996) produced breakfast cereals with 4.0, 6.0, and 8.4 g β-glucan respectively. The maximum increases observed in plasma glucose after eating the breakfast cereal were 67%, 42%, and 38% compared with the control. There was a linear inverse relationship between dose of β-glucan and plasma glucose peak or area under the glucose curve. Postprandial insulin increase was only 59% to 67% of that caused by control. A linear decrease in glycemic index was also found for increasing β-glucan content in bread (Cavallero et al., 2002).

The molecular weight of cereal β-glucan can be changed by β-glucanase. The results from Tosh et al. (2008) displayed that as the molecular weight was reduced from 2.2×10^6 to 4.0×10^5, the solubility of the β-glucan increased from a mean of 44% to 57%, but as the molecular weight was further decreased to 1.20×10^5, solubility fell to 26%. The changes in solubility is similar to the effect of molecular weight on the gel forming process. As mentioned before, low molecular weight β-glucans are more likely to form gels. It is obviously that molecular weight is not the only factor that influences the physiological effects. The physiological solubility, or the extend of solubilization from the food matrix in physiological state is the other factor that responsible for the effect in lowering blood glucose levels. The combined effects of molecular weight (M_w) and physiological solubility (C) was calculated as $\log(C\times M_w)$. Linear correlation was observed between the peak blood glucose rise (PBGR) and $\log(C\times M_w)$ by many researchers (Tosh et al., 2008; Wood et al., 2000). Tosh et al. (2008) also explained from the glycemic response curves that PBGR was a more sensitive measure than area under the curve (AUC) for comparison of test meals because it distinguished smaller differences in glycemic response. Frank et al. (2004) found that the molecular weight of oat β-glucan, which was incorporated into oat bran breads, may not play an important physiological role in moderately hypercholesterolemic humans. The result seems to contradict earlier conclusions. But from the experimental design we can find that the physiological solubility of β-glucan was neglected. The β-glucan in the bread was extracted with hot water instead of an *in vitro* extraction mimicking physiological state. Since the β-glucan fraction from hot water extraction is different from β-glucan soluble at physiological state. The determined data could not reflect the real situation in gastrointestinal tract.

Viscous soluble dietary fiber is useful in modifying postprandial hyperglycemia (Jenkins et al., 1978). The effect of viscosity on the physiological function of β-glucan has been tested recently. Oat β-glucan concentrations from an aqueous extraction process or an alcohol-based enzymatic process were added into an oral glucose drink with the same β-glucan content. Then the prepared drinks were consumed by subjects and blood glucose were analyzed. The incremental AUC of high viscosity sample was 19.6% and 17% lower than that of low viscosity sample and control, respectively (Panahi et al., 2007). The research from Juvonen et al. (2009) came to a similar conclusion that high viscosity of beverage containing β-glucan was more effective in lowering the blood glucose and insulin levels. The authors also found that β-glucan could influence the short-term gut hormone responses. The quantitative study on the relation between viscosity and postprandial glucose level was first carried out by Wood et al. (1994b). There was a highly significant linear relationship between log(viscosity) of the β-glucan beverage consumed and the glucose and insulin responses. The relationship showed that 79% to 96% of the changes in plasma glucose and insulin are attributable to viscosity. The authors deduced the relationship between viscosity and molecular weight and effective concentration later (Wood et al., 2000). If zero shear mass specific viscosity is used, then the PBGR can be related to concentration (C) and molecular weight (M_w) of β-glucan directly. There is a linear relationship between PBGR and $\log(C \times M_w)$. The relationship was varified in different kinds of oat-based foods recently (Regand et al., 2009). Although the molecular weight measurement was done on the original barley flour instead of on the bread β-glucan, Östman et al. (2006) still found a high correlation ($r^2 = 0.96$; $P = 0.0007$) between the fluidity index (FI) of the enzymatic digests and glycemic index (GI) of bread products ($GI = 50.8 + 0.441FI$).

Processing of food containing β-glucan may change the molecular weight or solubility, hence cause different physiological effects in different food matrix. Drink model is simpler than solid food model because molecular weight and physiological solubility experience more changes in solid foods than in liquid beverages. The influence of differently processed oat foods on glycemic response were studied by Regand et al. (2009). The results showed that porridge and granola had the highest efficacy in attenuating the PBGR because of their high peak molecular weight and viscosity. Cereal β-glucan depolymerization in bread and pasta reduced β-glucan bioactivity. Casiraghi et al. (2006) pointed out that barley β-glucan added in cookies was more effective than that added in crackers. As mentioned before, cereal β-glucans may be degraded by enzymes in bread making process. Freeze-thaw cycles tend to decrease the solubility of β-glucan. Food processing can increase the physiological activity of β-glucans by increasing availability (cooking, extrusion), while the molecular weight of the polymer may be reduced. Since processing has a great impact on the physiological function of β-glucan. The unfavorable processing should be avoided if the aim is to retain the physiological effects.

7.2. Effect on Plasma Cholesterol

Cholesterol-lowering effect of oat based food may first be proposed by De Groot et al. (1963). About 30 years later, the soluble fiber in oat was considered as the effective factor that could lower the blood lipid (Ripsin et al., 1992). Then Braaten et al. (1994) pointed out that β-glucan was responsible for the cholesterol-lowering effect of oat. But the process by which cereal β-glucans lower serum cholesterol is not completely understood till now. There is evidence of two probable mechanisms. One of the hypothesis emphasises the importance of viscosity formed by β-glucans in small intestine. The other hypothesis focuses on the effect of the fermentation products of β-glucans in the gut, which may alter cholesterol synthesis. Addition of cereal β-glucans into the feed increased the viscosity of small intestinal digesta in rats and chicks (Danielson et al., 1997; Wang et al., 1992). The viscosity decreased the digestion and absorption of lipids and resulted lower serum total and low density lipoprotein (LDL) cholesterol levels. The formation of a thick unstirred water layer, adjacent to the mucosa may act as a physical barrier to reduce the absorption of nutrients and bile acids. Viscosity may reduce the rate of glucose absorption, leading to a lower glycemic response and lower insulin concentrations, hence result in a reduced hepatic cholesterol synthesis (Bell et al., 1999). Cereal β-glucans may increase the binding of bile acids in the intestinal tract, leading to a decreased enterohepatic circulation of bile acids and a subsequent increase in the hepatic conversion of cholesterol to bile acids (Glore et al., 1994; Bell et al., 1999). Since cereal β-glucans are able to interact with steroids and to transport greater amounts of bile acid and neutral sterols towards lower parts of the intestinal tract. The greater excretion of bile acid and neutral sterols in the presence of the higher contents of barley fiber may be important for the reduction of lipids in the serum of hypercholesterolemic individuals (Dongowski et al., 2003). Cereal β-glucans inhibited the *in vitro* intestinal uptake of long chain fatty acids and cholesterol, and further reduced intestinal fatty acid binding protein and fatty acid transport protein 4 mRNA. The expression of genes involved in fatty acid synthesis and cholesterol metabolism was down regulated with the β-glucan extracts (Drozdowski et al., 2010). The viscosity and the binding ability are direct effects on cholestetol. In addition, cereal β-glucans can be fermented by microorganisms in human gut, the main product is SCFA. It has been shown in animals that production of SCFA, such as acetate, propionate, and butyrate, inhibited hepatic cholesterol synthesis (Bridges et al., 1992; Wright et al., 1990; Queenan et al., 2007).

As far as viscosity is concerned, the effects of dose (more specificly, physiological solubility or availability), molecular weight, and viscosity should be evaluated. Untreated oat bran and enzyme hydrolyzed oat bran were incorporated into diets and fed to rats for 4 weeks (Tietyen et al., 1995). The results showed that hepatic cholesterol accumulation was less in the oat bran group than in the enzyme-treated oat bran or cellulose groups. Davidson et al. (1991) reported that 56 g of oat bran resulted in significantly greater reductions in LDL cholesterol levels than 56 g of oatmeal. Since the content of oat β-glucan is usually higher in bran than in oatmeal, the results indicate that higher β-glucan content is more effective. This is consistent with viscosity hypothesis. But the results also showed that 56 g and 84 g oat bran decreased LDL cholesterol levels by 15.9% and 11.5% rspectively. It seemes that lower β-glucan content is more effective. The results was confusing. Yokoyama et al. (1998) found that Oatrim, a product developed for use as a fat substitute, could also reduce plasma cholesterol due to its soluble dietary fiber content. But it is surprised that Oatrim hydrolysate (0% β-glucan) reduced the LDL cholesterol levels in animals. The authors hypothesized that other components such as phytosterols may be responsible for the cholesterol lowering. Different molecular weight oat β-glucans were used to test the effect of molecular weight on the blood lipid of mice (Bae et al., 2009). The results showed that food samples containing β-glucan lowered the serum lipid significantly. However, the serum lipid profile between samples containing β-glucans with different molecular weight was similar, indicating that molecular weight was not the only critical factor for controlling lipid profile. The results from another nutrition research displayed that molecular weight of the barley β-glucan did not alter the serum lipid levels significantly in mildly hypercholesterolemic adults (Smith et al., 2008). On the contrary, Wilson et al. (2004) found that decreases in plasma total cholesterol and non-HDL (high density lipoprotein) cholesterol concentrations occurred in the hamsters fed reduced molecular weight and high molecular weight β-glucan diets. Plasma HDL cholesterol concentration did not differ. Fecal excretion of cholesterol and sterols was also increased by β-glucan. The above results are confusing probably due to other factors that were not controlled in the experiments.

Viscosity is a suitable parameter to describe the relation between hypocholesterolemic effect and physicochemical property of cereal β-glucans. When chicks were fed a cornsoybean meal diet, a barley diet with β-glucanase and that diet without β-glucanase. Significant ($P < 0.01$) negative correlations occurred between viscosity of the small intestinal contents and plasma total and LDL cholesterol concentrations (Wang et al., 1992). Lipid levels in rats were lowered when the rats were fed with diets containing barley β-glucan (Danielson et al., 1997). The authors found that viscosity of intestinal contents increased significantly ($P < 0.05$) when rats were fed increasing levels of β-glucan. Recently, the effect of physicochemical properties of oat β-glucan on cholesterol-lowering ability was evaluated in humans (Wolever et al., 2010). There was a negatively linear relationship between LDL cholesterol concentration at 4 week and log(viscosity) or log($M_W \times C$). Which is similar to the relationship in blood glucose lowering effect. Just like the effect of β-glucan on lowering blood glucose, food matrix may have effect on the physicochemical state of β-glucan and lead to high or low effect. It showed that oat β-glucan was more effective in lowering cholesterol when incorporated in orange juice than in bread and cookies (Kerckhoffs et al., 2003). Food processing may increase the availability of β-glucan, whereas the molecular weight usually decreases.

Bile acid binding capacity of cereal β-glucans is supposed to be one of the mechanisms in hypocholesterolemic effects. But the relation between the effect and the content or molecular weight of β-glucan is not clear. The results from several experiments using cereal bran or flour showed that bile acid binding was not related to β-glucan contents. The insoluble dietary fiber such as lignin was responsible for the bile acid binding capacity (Kahlon and Wooddruff, 2003; Sayar et al., 2005, 2006). There was no significant correlation between the molecular weight and the bile acid binding (Bae et al., 2009). Oat β-glucan was treated with lichenase to yield high, medium, and low molecular weight fractions (Kim and White, 2010). The results showed that lower molecular weight oat β-glucan had higher binding capacity for bile acid. Probably the lower viscosity of low molecular weight β-glucan is favourable to the diffusion of bile acid and hence the combination between β-glucan and bile acid.

Another possibility is that increased solubility of low molecular weight β-glucan may enhance the bile acid binding. Increasing the solubility of β-glucan can be achieved by derivation. 2,2,6,6-tetramethyl-1-piperidine oxoammonium ion (TEMPO) was used for oxidation of oat β-glucan. The oxidized oat β-

glucan was effective in lowering serum lipids due to enhanced water solubility and improved *in vitro* bile acid binding capacity (Park et al., 2009).

Under the mildest acid conditions, 37 °C and pH 1, corresponding to those in human stomach, no degradation of β-glucan was observed with HCl over a 12 h period (Johansson et al., 2006). Whereas cereal β-glucans can be fermented in human and animal gut by microbes. Studying fermentation of cereal β-glucan in human gut is difficult due to the invasive and expensive nature of colonic observation and the dynamic nature of the colon.

Excreted colon contents will not necessary represent colon contents, due to continual fermentation of fiber and absorption of minerals and SCFA across the epithelium. *In vitro* fermentation is a noninvasive, time-efficient means to estimate fiber fermentability. The fermentation product, SCFA, especially propionate can inhibit the synthesis of cholesterol (Wolever et al., 1991). *In vitro* fermentation of oat flours from typical and high β-glucan oat lines showed that more propionate and butyrate, but less acetate, were produced from high-β-glucan oat flours (Sayar et al., 2007). The ropionate and butyrate are helpful in lowering cholesterol. Six grams oat β-glucan per day for six weeks reduced total and LDL cholesterol levels in human subjects significantly (Queenan et al., 2007).

Base on an *in vitro* intestinal fermentation model, the authors found that fermentation of oat β-glucan produced total SCFA and acetate concentrations similar to inulin and guar gum. Whereas oat β-glucan produced the highest concentrations of butyrate at 4, 8, and 12 hours. Kim and White (2010) found that the low molecular weight β-glucan produced greater amounts of SCFA than the high molecular weight β-glucan after 24 h of fermentation. Among the major SCFA, more propionate was produced from all the oat β-glucan samples.

In another nutrition study, pigs fed diets with oat β-glucan had a higher ($P < 0.05$) net absorption of SCFA than pigs fed control diets without oat β-glucan, indicating increased fermentation in the gut of pigs (Hooda et al., 2010).

7.3. Immune Modulation

Immune activity of β-glucans from yeast and fungi have been studied extensively. While the reports on the immune effects of cereal β-glucans are rare. Because cereal β-glucans exist in human diet, they contact with enterocytes directly. Its reasonable that cereal β-glucans may adjust immune system in gut. Ramakers et al. (2007) reported that fecal water prepared from ileostomy contents of patients who had consumed diet containing oat β-glucan increased the immune response of *in vitro* cultured enterocyte cell lines. But the immune effects were not mediated by β-glucan receptor dectin-1 (Volman et al., 2010a). The authors carried out another experiment and found that oat β-glucan increased NF-κB transactivation of intestinal leukocytes and enterocytes in the proximal part of the small intestine in mice. The elevated NF-κB activation may help to prevent infection with pathogens. Since there was an elevated level of IL-12 in intestine, dietary oat β-glucan may activate intestinal leukocytes, which in turn induce a mild inflammatory phenotype of enterocytes (Volman et al., 2010b). Wang et al. (2010) provided evidence that whole oat significantly inhibits aberrant crypt foci (ACF) formation and colon tumor growth in mice. Which suggested that whole oat could be a good food source for prevention of colon cancer. Oat β-glucan may be involved in the anticancer effect by activating gut-associated lymphoid tissue (GALT). Another possible pathway is related to the fermentation products of oat β-glucan. The produced SCFA, especially butyrate, are effective in preventing colon cancer as well as providing energy for intestinal cells. Of course, the anticancer effect of oat can not be ascribed to β-glucan completely. Vitamins, antioxidants, phenolic compounds and minerals are important nutrients having potent anticancer activity.

Cereal β-glucans can provide resistance to infectious diseases and parasites in amimals. Yun et al. (1997) came to a conclusion that immune functions suppressed by DXM, an immunosuppressant, could be partially restored by oat β-glucan. The restored immune functions may increase the resistance to coccidial infection. The research from the same group indicated that oat β-glucan did not influence immune responses of naive cells *in vitro* or of healthy steers *in vivo*; however, when cells or animals were treated with DXM, oat β-glucan significantly restored some of the immune parameters studied (Estrada et al., 1999). The oral or parenteral oat β-glucan treatment also enhanced the resistance to *S. aureus* and *E. vermiformis* infection in the mice (Yun et al., 2003). Moderate exercise and oat β-glucan were evaluated for their functioin on immune activation and resistance to respiratory infection in mice. The obtained data confirmed a positive effect of both moderate exercise and oat β-glucan on immune function, but only moderate exercise was associated with a significant reduction in the risk of upper respiratory tract infection (Davis et al., 2004). Another experiment from the same research group indicated that oat β-glucan can decrease the metatastic spread of injected B16 melanoma cells in animals, and these effects may be mediated in part by an increase in macrophage cytotoxicity to B16 melanoma (Murphy et al., 2004). Effect of molecular weight of cereal β-glucan on the immune function was studied by Roubroeks et al. (2000c). Ray β-glucan fractions with molecular weight between 7.98×10^4 and 1.39×10^4 were used for stimulation of human monocytes to produce tumour necrosis factor (TNF). The fration with a molecular weight of 1.89×10^4 was found to be the most potent immunostimulator.

7.4. Other Functions

Except the activity in lowering blood glucose and serum lipids, cereal β-glucans showed other potential physiological benefits. Since sulfation of polysaccharides may increase the solubility and enhance biological activities. Oat β-glucan was sulfated and the product displayed anti-HIV activity by preventing entry of the virus to the cell (Wang et al., 2008). Rats were fed diets with or without cereal β-glucans. The results showed that lactobacilli formed a greater proportion of the gut microbes in β-glucan-fed rats. *In vitro* experiments confirmed that some lactobacilli could utilize oligosaccharides (DP3 and DP4) in β-glucan hydrolysates (Snart et al., 2006). The result indicated that β-glucan can produce a prebiotic effect in the gut of rats. It is interesting that oat β-glucan has the ability to penetrate the skin into the epidermis and dermis, which showed a significant reduction of wrinkle (Pillai et al., 2005).

8. Application in Food Industry

Addition of cereal β-glucan into food system will not only improve the sensory quality of the products, but also bring beneficial effects to consumers. Cereal β-glucan is an important ingredient for food industry due to the unique rheological behavior and gelling property. It can be uesd as a natural hydrogel in food. Since cereal β-glucan is capable of forming a membrane, it has the potential to be made into a biodegradable package material for food (Skendi et al., 2003). Commercial β-glucan ingredients include "trim" family like Oatrim, Nutrim, C-trim (Calorie-trim), and Z-trim.

Cereal β-glcuans have been successfully incorporated into pasta as a functional ingredient. Yokoyama et al. (1997) compared blood glucose level and insulin response of healthy volunteers given a control durum wheat pasta and a pasta sample with added barley β-glucan. The results showed that the blood glucose and insulin response were significantly reduced by pasta added β-glucan. Cereal β-glucans have also been used in making bread. Cavallero et al. (2002) added ingredient rich in barley β-glucan into wheat bread. The results showed that a linear decrease in glycemic index was associated with increasing β-glucan concentrations. Which was ascribed to the effect of β-glucan on digesta viscosity and glucose absorption. Brennan and Cleary (2007) enriched wheat bread with 5% gelling barley β-glucan (Glucagel) and the bread significantly reduced the rate at which reducing sugars were released in an *in vitro* digestion model. The results also indicated that lower additions (2.5%) of the β-glucan material could increase the potential glycemic index of bread, probably by increasing starch digestibility. The dose dependence needs further research and explanation.

A lot of fat replacement products have been developed from β-glucan concentrate. Inglett (1993) prepared amylodextrins containing 10% β-glucan (Oatrim) from oat flours and bran by α-amylase, the product can be used as fat substitute. The mixture containing 10% or 24% β-glucan displayed shear-thickening behavior probably due to complicated interactions between the β-glucan and amylodextrins (Carrierea and Inglett, 1999). Cocoa butter in chocolates was replaced by C-trim30 with reduced calories. The use of C-trim30 up to 10% produced softer chocolates. Since the viscosity of chocolate dramatically increased with the addition of C-trim30, only a limited quantity could be used to replace the cocoa butter (Lee et al., 2009a). Oat fiber in gel form (Nutrim-10) was added at 13.4% to low-fat beef patties. The ingredient increased the cooking yield, retentions of fat, and moisture. It also maintained the tenderness of the product by enhanceing moisture and fat entrapment without perceivable effect on sense and storage quality (Piñero et al., 2008). Oatrim was utilized as replacer for shortening in cakes and reduced the calories intake. The product did not show significant difference form the control (Lee et al., 2005). Z-trim has been added to beef patties. Sensory analysis of flavor and texture of the products showed slight decreases in beef and fatty flavors and increases in texture characteristics of tenderness and juiciness (Warner and Inglett, 1997). In making oatmeal raisin cookies, oatrim gel can be substituted for shortening to 50% level without significant effect on flavor and texture of the cookie. Oatrim can also replace whipping cream in truffle centers without causing significant flavor intensities at the 50% cream level (Inglett et al., 1994).

Cereal β-glucan has been used as a emulsifying stabilizer in food system. The stability of egg-yolk-stabilized model emulsions containg β-glucan was investigated by Kontogiorgos et al. (2004). The authors concluded that the high molecular weight β-glucans stabilized emulsions mainly by increasing the viscosity of continuous phase, while the stability effect of low molecular weight β-glucans was mainly caused by network formation in the continuous phase. The structural variation, such as the ratio of DP3 to DP4, had weak impact on the stability of the emulsion for hight molecular weight β-glucans. Whereas the low molecular weight β-glucans with different fine structure showed different rheological behavior in the system. Burkus and Temelli (2000) found that when whey protein concentrate was used as a emulsifier and foaming agent, addition of barley β-glucan increased the stability of emulsions and foams. It was interesting that sugar acted synergistically with high-viscosity barley β-glucan, leading to increased stability of the above system. The main mechanism for increased stability was ascribed to increased viscosity by β-glucan. The milk protein products are used as functional ingredients in many food. Addition of β-glucan to food systems containing sodium caseinate or whey protein may lead to phase separation and influence the stability and functional properties of food matrix.

The phase separation behavior of β-glucan and milk protein mixtures has been studied. Lazaridou and Biliaderis (2009) found that the phase separation depended on molecular weight of oat β-glucan. The low molecular weight β-glucan tended to form gel and promoted phase separation. The phase separation kinetics of oat β-glucan and whey protein mixtures were investigated by Kontogiorgos et al. (2009a, b).

Cereal β-glucans are functional addictives for improving food quality. C-trim30, a hydrocolloid rich in oat β-glucan, was added in batters for fried foods. The oat β-glucan increased the viscosity and pickup of batters, while the loss of moisture was decreased. The hydrogel also reduced oil take up of fried foods (Lee and Inglett, 2007). The addition of β-glucan into bread resulted in increased bread loaf volume at 1% and 2% concentration. The porosity of crumb and lightness of crust were elevated at 1% concentration. While 2% β-glucan significantly increased crumb firmness (Lazaridou et al., 2007). Nutrim-OB and C-trim20 could partially replace wheat flour to 20% and 10% respectively without negative effect on flavor and texture. In making peanut spreads, Nutrim-OB could partially replace peanut flour up to 13% and oil up to 5% without undesirable changes. It also lowered the calorie intake (Lee et al., 2009b). The cookies containing C-trim20 exhibited increased elastic properties and water content with the similar texture and sensory as control.

β-glucans are suitable for dairy industry. Low-fat ice cream and yoghurt have been produced with β-glucan (Brennan and Tudorica, 2008). Addition of

β-glucan into dairy products can improve the sensory property and give the products similar sensory as full-fat products. Using barley β-glucan in milk systems will promote the coagulation of the milk and improve the texture of the fresh curds (Tudorica et al., 2004). But sometimes β-glucan fortified dairy product, such as white-brined cheese, is inferior in the colour, flavour and sensory to those of a typical white-brined cheese product (Volikakis et al., 2004).

REFERENCES

Ajithkumar, A.; Andersson, R.; Siika-aho, M.; Tenkanen, M.; Åman, P. Isolation of cellotriosyl blocks from barley β-glucan with endo-1,4-β-glucanase from *Trichoderma reesei*. *Carbohydrate Polymers*, 2006, *64*, 233–238.

Åman, P.; Rimsten, L.; Andersson, R. Molecular weight distribution of β-glucan in oat-based foods. *Cereal Chemistry*, 2004, *81*, 356–360.

Annison, G. Relationship between the levels of soluble nonstarch polysaccharides and the apparent metabolizable energy of wheats assayed in broiler chickens. *Journal of Agricultural and Food Chemistry*, 1991, *39*, 1252–1256.

Autio, K.; Salmenkallio-Marttila, M. Light microscopic investigations of cereal grains, doughs and breads. *Lebensmittel-Wissenschaft und-Technologie*, 2001, *34*, 18–22.

Bae, I. Y.; Lee, S.; Kim, S. M.; Lee, H. G. Effect of partially hydrolyzed oat b-glucan on the weight gain and lipid profile of mice. *Food Hydrocolloids*, 2009, *23*, 2016–2021.

Beer, M. U.; Arrigoni, E.; Amado, R. Extraction of oat gum from oat bran: effects of process on yield, molecular weight distribution, viscosity and (1→3)(1→4)-β-D-glucan content of the gum. *Cereal Chemistry*, 1996, *73*, 58–62.

Beer, M. U.; Wood, P. J.; Weisz, J. Molecular weight distribution and (1→3)(1→4)-β-D-glucan content of consecutive extracts of various oat and barley cultivars. *Cereal Chemistry*, 1997a, *74*, 476–480.

Beer, M. U.; Wood, P. J.; Weisz, J.; Filliion, N. Effect of cooking and storage on the amount and molecular weight of (1→3)(1→4)-β-D-glucan extracted from oat products by an *in vitro* digestion system. *Cereal Chemistry*, 1997b, *74*, 705–709.

Bell, S.; Goldman, V. M.; Bistrian, B. R.; Arnold, A. H.; Ostroff, G.; Forse, R. A. Effect of β-glucan from oats and yeast on serum lipids. *Critical Reviews in Food Science and Nutrition*, 1999, *39*, 189–202.

Bergh, M. O.; Razdan, A.; Aman, P. Nutritional influence of broiler chicken diets based on covered normal, waxy and high amylose barleys with or without enzyme supplementation. *Animal Feed Science and Technology*, 1999, *78*, 215–226.

Bhatty, R. S. Total and extractable β-glucan contents of oats and their relationship to viscosity. *Journal of Cereal Science*, 1992, *15*, 185–192.

Bhatty, R. S. Laboratory and pilot plant extraction and purification of β-glucans from hull-less barley and oat brans. *Journal of Cereal Science*, 1995, *2*, 163–170.

Bock, K.; Duus, J. O.; Norman, B.; Pedersen, S. Assignment of structures to oligosaccharides produced by enzymic degradation of a β-D-glucan from barley by ^{1}H- and ^{13}C-n.m.r. spectroscopy. *Carbohydrate Research*, 1991, *211*, 219–233.

Böhm, N.; Kulicke, W. M. Rheological studies of barley (1→3), (1→4)-β-glucan in concentrated solution: investigation of the viscoelastic flow behaviour in the sol-state. *Carbohydrate Research*, 1999a, *315*, 293–301.

Böhm, N.; Kulicke, W. M. Rheological studies of barley (1→3), (1→4)-β-glucan in concentrated solution: mechanistic and kinetic investigation of the gel formation. *Carbohydrate Research*, 1999b, *315*, 302–311.

Braaten, J. T.; Wood, P. J.; Scott, F. W.; Wolynetz, M. S.; Lowe, M. K.; Bradley-White, P.; Collins, M. W. Oat β-glucan reduces blood cholesterol concentration in hypercholesterolemic subjects. *European Journal of Clinical Nutrition*, 1994, *48*, 465–474.

Brennan, C. S.; Cleary, L. J. Utilisation Glucagel® in the β-glucan enrichment of breads: A physicochemical and nutritional evaluation. *Food Research International*, 2007, *40*, 291–296.

Brennan, C. S.; Tudorica, C. M. Carbohydrate-based fat replacers in the modification of the rheological, textural and sensory quality of yoghurt: comparative study of the utilisation of barley beta-glucan, guar gum and inulin. *International Journal of Food Science and Technology*, 2008, *43*, 824–833.

Bridges, S. R.; Anderson, J. W.; Deakins, D. A.; Dillon, D. W.; Wood, C. L. Oat bran increases serum acetate of hypercholesterolemic men. *American Journal of Clinical Nutrition*. 1992, *56*, 455–459.

Burkus, Z. Temelli, F. Stabilization of emulsions and foams using barley *β*-glucan. *Food Research International*, 2000, *33*, 27–33.

Carr, J. M; Glatter, S.; Jeraci, J. L.; Lewis, B. A. Enzymic determination of *β*-glucan in cereal-based food products. *Cereal Chemistry*, 1990, *67*, 226–229.

Carrierea, C. J.; Inglett, G. E. Nonlinear viscoelastic solution properties of oat-based *β*-glucan/amylodextrin blends. *Carbohydrate Polymers*, 1999, *40*, 9–16.

Casiraghi, M. C.; Garsetti, M.; Testolin, G.; Brighenti, F. Post-prandial responses to cereal products enriched with barley β-glucan. *Journal of the American College of Nutrition*, 2006, *25*, 313–320.

Cavallero, A.; Empilli, S.; Brighenti, F.; Stanca, A. M. High (1→3, 1→4)-*β*-D-glucan fractions in bread making and their effect on human glycaemic response. *Journal of Cereal Science*, 2002, *36*, 59–66.

Colleoni-Sirghiea, M.; Fultonb, D. B.; White P. J. Structural features of water soluble (1,3) (1,4)-*β*-D-glucans from high-*β*-glucan and traditional oat lines. *Carbohydrate Polymers*, 2003, *54*, 237–249.

Cui, W.; Wood, P. J.; Blackwell, B.; Nikiforuk J. Physicochemical properties and structural characterization by two-dimensional NMR spectroscopy of wheat *β*-D-glucan—comparison with other cereal *β*-D-glucans. *Carbohydrate Polymers*, 2000, *41*, 249–258.

Dais, P.; Perlin, A. S. High-field 13C-N.M.R. spectroscopy of *β*-Dglucans, amylopectin, and glycogen. *Carbohydrate Research*, 1982, *100*, 103–116.

Danielson, A. D.; Newman, R. K.; Newman, C. W.; Berardinelli, J. G. Lipid levels and digesta viscosity of rats fed a high-fiber barley milling fraction. *Nutrition Research*, 1997, *17*, 515–522.

Davidson, M. H.; Dugan, L. D.; Burns, J. H.; Bova, J.; Story, K.; Drennan, K. B. The hypocholesterolemic effects of beta-glucan in oatmeal and oat bran. A dose-controlled study. *Journal of the American Medical Association*, 1991, *265*, 1833–1839.

Davies, L. M.; Harris, P. J.; Newman, R. H. Molecular ordering of cellulose after extraction of polysaccharides from primary cell walls of Arabidopsis thaliana: A solid-state CP/MAS ^{13}C NMR study. *Carbohydrate Research*, 2002, *337*, 587–593.

Davis, J. M.; Murphy, E. A.; Brown, A. S.; Carmichael, M. D.; Ghaffar, A.; Mayer, E. P. *American Journal of Physiology - Regulatory, Integrative and Comparative Physiology*, 2004, *286*, 366–372.

De Groot, A. P.; Luyken, R.; Pikaar, N. A. Cholesterol-lowering effect of rolled oats. *Lancet*, 1963, *2*, 303–304.

Degutyte-Fomins, L.; Sontag-Strohm, T.; Salovaara, H. Oat bran fermentation by rye sourdough. *Cereal Chemistry*, 2002, *79*, 345–348.

Doehlert, D. C.; Moore, W. R. Composition of oat bran and flour prepared by three different mechanisms of dry milling. *Cereal Chemistry*, 1997, *74*, 403–406.

Dongowski, G.; Huth, M.; Gebhardt, Erich. Steroids in the intestinal tract of rats are affected by dietary-fibre-rich barley-based diets. *British Journal of Nutrition*, 2003, *90*, 895–906.

Doublier, J. L.; Wood, P. J. Rheological properties of aqueous solutions (1→3)(1→4)-β-D-glucan from oats (*Avena sativa* L.). *Cereal Chemistry*, 1995, *72*, 335–340.

Drozdowski, L. A.; Reimer, R. A.; Temelli, F; Bell, R. C.; Vasanthan, T; Thomson, A. B. R. β-Glucan extracts inhibit the *in vitro* intestinal uptake of long-chain fatty acids and cholesterol and down-regulate genes involved in lipogenesis and lipid transport in rats. *Journal of Nutritional Biochemistry*, 2010, *21*, 695–701.

Estrada, A.; Van Kessel, A.; Laarveld, B. Effect of administration of oat β-glucan on immune parameters of healthy and immunosuppressed beef steers. *Canadian Journal of Veterinary Research*, 1999, *63*, 261–268.

FDA. Food labeling: health claims; oats and coronary heart disease. *Federal Register*, 1997, *62*, 3583–3601.

FDA. Food labeling: health claims; soluble dietary fiber from certain foods and coronary heart disease. *Federal Register*, 2005, *70*, 76150–76162.

Frank, J.; Sundberg, B.; Kamal-Eldin, A.; Vessby, B.; Åman, P. Yeast-leavened oat breads with high or low molecular weight β-glucan do not differ in their effects on blood concentrations of lipids, insulin, or glucose in humans. *Journal of Nutrition*. 2004, *34*, 1384–1388.

Freijee, F. J. M. Determination of the high molecular weight β-glucan content of malt wort by a spectrophotometric method—Determination of the accuracy, repeatability and reproducibility. *Journal of the Institute of Brewing*, 2005., *111*, 341–343.

Fulcher, G. B.; Miller, S. S. Structure of oat bran and distribution of dietary fiber components, oat bran. In: Wood, P. J. ed., *Cereal Chemistry*. St Paul, MN: American Association of Cereal Chemists, 1993, 1–24.

Glore, S. R.; Van Treeck, D.; Knehans, A. W.; Guild, M. Soluble fiber and serum lipids: a literature review. *Journal of American Dietary Association*, 1994, *94*: 425–436.

Gómez, C; Navarro, A.; Manzanares, P.; Hortab, A.; Carbonell, J. V. Physical and structural properties of barley (1→3),(1→4)-*β*-D-glucan. Part I. Determination of molecular weight and macromolecular radius by light scattering. *Carbohydrate Polymers*, 1997a, *32*, 7–15.

Gómez, C; Navarro, A.; Manzanares, P.; Hortab, A.; Carbonell, J. V. Physical and structural properties of barley (1→3),(1→4)-*β*-D-glucan. Part II. Viscosity, chain stiffness and macromolecular dimensions. *Carbohydrate Polymers*, 1997b, *32*, 17–22.

Gómez, C; Navarro, A.; Manzanares, P.; Hortab, A.; Carbonell, J. V. Physical and structural properties of barley (1→3),(1→4)-*β*-D-glucan. Part III. Formation of aggregates analysed through its viscoelastic and flow behaviour. *Carbohydrate Polymers*, 1997c, *34*, 141–148.

Grimm, A.; Kruger, E.; Burchard, W. Solution properties of *β*-D-(1,3)(1,4)-glucan isolated from beer. *Carbohydrate Polymers*, 1995, *27*, 205–214.

Guillon, F.; Champ, M. Structural and physical properties of dietary fibres, and consequences of processing on human physiology. *Food Research Intenational*, 2000, *33*, 233–245.

Haigler, C. H.; Brown, R. M.; Benziman, M. Calcofluor white ST alters the in vivo assembly of cellulose microfibrils. *Science*, 1980, *210*, 903–906.

Harding, S. E.; Vårum, K. M.; Stokke, B.T.; Smidsrød, O. Molecular weight determination of polysaccharides. *Advances in Carbohydrate Analysis*, 1991, *1*, 63–144.

Henriksson, K.; Telemad, A.; Suorttia, T.; Reinikainen, T.; Jaskarib, J.; Teleman, O.; Poutanen, K. Hydrolysis of barley (1→3), (1→4)-*β*-D-glucan by a cellobiohydrolase II preparation from *Trichoderma reesei*. *Carbohydrate Polymers*, 1995, *26*, 109–119.

Hlebowicz, J.; Darwiche, G.; Björgell, O.; Almér, L.-O. Effect of muesli with 4 g oat β-glucan on postprandial blood glucose, gastric emptying and satiety in healthy subjects: a randomized crossover trial. *Journal of the American College of Nutrition*, 2008, *27*, 470–475.

Hooda, S.; Matte, J. J.; Vasanthan, T.; Zijlstra, R. T. Dietary oat β-glucan reduces peak net glucose flux and insulin production and modulates plasma incretin in portal-vein catheterized grower pigs. *Journal of Nutrition*, 2010, *140*, 1564–1569.

Hu, X.; Xing, X.; Ren, C. The effects of steaming and roasting treatments on *β*-glucan, lipid and starch in the kernels of naked oat (*Avena nuda*). *Journal of the Science of Food and Agriculture*, 2010, *90*, 690–695.

Hübner, F.; O'Neil, T.; Cashman, K. D.; Arendt, E. K. The influence of germination conditions on beta-glucan, dietary fibre and phytate during the germination of oats and barley. *European Food Research and Technology*, 2010, *231*, 27–35.

Inglett, G. E. Amylodextrins containing β-glucan from oat flours and bran. *Food Chemistry*, 1993, *47*, 133–136.

Inglett, G. E.; Warner, K.; Newman, R. K. Sensory and nutritional evaluations of oatrim. *Cereal Foods World*, 1994, *39*, 755-756, 758–759.

Irakli, M.; Biliaderis, C. G.; Izydorczyk, M. S.; Papadoyannis, I. N. Isolation, structural features and rheological properties of water-extractable β-glucans from different Greek barley cultivars. *Journal of the Science of Food and Agriculture*, 2004, *84*, 1170–1178.

Izawa, M.; Takashio, M.; Tamaki, T. Inhibitor of fluorescence reactions between calcofluor and β-(1→3),(1→4)-D-glucan in beer and wort. *Journal of the Institute of Brewing*, 1996, *102*, 87–91.

Izydorczyk, M. S.; Biliaderis, C. G.; Macri, L. J.; MacGregor, A. W. Fractionation of oat (1→3), (1→4)-β-D-glucans and characterisation of the fractions. *Journal of Cereal Science*, 1998a, *27*, 321–325.

Izydorczyk, M. S; Macri, L. J.; MacGregor, A. W. Structure and physicochemical properties of barley non-starch polysaccharides - I. Waterextractable β-glucans and arabinoxylans. *Carbohydrate Polymers*, 1998b, *35*, 249–258.

Izydorczyk, M. S; Storsley, J.; Labossiere, D.; MacGregor, A. W.; Rossnagel, B. G. Variation in total and soluble β-glucan content in hulless barley: effects of thermal, physical, and enzymic treatments. *Journal of Agricultural and Food Chemistry*, 2000, *48*, 982–989.

Jaskari, J.; Henriksson, K.; Nieminen, A.; Suortti, T.; Salovaara, H.; Poutanen, K. Effect of hydrothermal and enzymic treatments on the viscous behavior of dry- and wet-milled oat brans. *Cereal Chemistry*, 1995, *7*, 625–631.

Jenkins, D .J. A.; Wolever, T. M. S.; Leeds, A. R.; Gassull, M. A.; Dilawari, J. B.; Goff, D. V.; Metz, G. L.; Alberti, K. G. M. Dietary fibres, fibre analogues and glucose tolerance, importance of viscosity. *British Medical Journal*, 1978, *1*, 1392–1394.

Jeroch, H.; Danicke, S. Barley in poultry feeding: a review. *World's Poultry Science Journal*, 1995, *51*, 271–291.

Jiang, G.; Vasanthan, T. MALDI-MS and HPLC quantification of oligosaccharides of lichenase-hydrolyzed water-soluble β-glucan from ten barley varieties. *Journal of Agricultural and Food Chemistry*, 2000, *48*, 3305–3310.

Johansson, L.; Virkkia, L.; Maunub, S.; Lehtoa, M.; Ekholma, P; Varoa, P. Structural characterization of water soluble β-glucan of oat bran. *Carbohydrate Polymers*, 2000, *42*, 143–148.

Johansson, L.; Tuomainen, P.; Ylinen, M.; Ekholm, P.; Virkki, L. Structural analysis of water-soluble and -insoluble β-glucans of whole-grain oats and barley. *Carbohydrate Polymers*, 2004, *58*, 267–274.

Johansson, L.; Virkki, L.; Anttila, H.; Esselström, H.; Tuomainen, P.; Sontag-Strohm, T. Hydrolysis of β-glucan. *Food Chemistry*, 2006, *97*, 71–79.

Johansson, L.; Tuomainen, P.; Anttila, H; Rita, H.; Virkki, L. Effect of processing on the extractability of oat β-glucan. *Food Chemistry*, 2007, *105*, 1439–1445.

Johansson, L.; Karesoja, M.; Ekholm, P.; Virkki, L.; Tenhu, H. Comparison of the solution properties of (1→3),(1→4)-β-D-glucans extracted from oats and barley. *LWT - Food Science and Technology*, 2008, *41*, 180–184.

Juvonen, K. R.; Purhonen, A.-K.; Salmenkallio-Marttila, M.; Lähteenmäki, L.; Laaksonen, D. E.; Herzig, K.-H.; Uusitupa, M. I. J.; Poutanen, K. S.; Karhunen, L, J. Viscosity of oat bran-enriched beverages influences gastrointestinal hormonal responses in healthy humans. *Journal of Nutrition*, 2009, *139*, 461–466.

Kahlon, T. S.; Woodruff, C. L. In vitro binding of bile acids by rice bran, oat bran, barley and β-glucan enriched barley. *Cereal Chemistry*, 2003, *80*, 260–263.

Karmally, W.; Montez, M. G.; Palmas, W.; Martinez, W.; Branstetter, A.; Ramakrishnan, R.; Holleran, S. F.; Haffner, S. M.; Ginsberg, H. N. Cholesterol-lowering benefits of oat-containing cereal in Hispanic Americans. *Journal of the American Dietetic Association*, 2005, *105*, 967–970.

Kerckhoffs, D. A.; Hornstra, G.; Mensink, R. P. Cholesterol lowering effect of β-glucan from oat bran in mildly hypercholesterolemic subjects may decrease when β-glucan is incorporated into bread and cookies. *American Journal of Clinical Nutriton*, 2003, *78*, 221–227.

Kim, H. J.; White, P. J. *In vitro* bile-acid binding and fermentation of high, medium, and low molecular weight β-glucan. *Journal of Agricultural and Food Chemistry*, 2010, *58*, 628–634.

Kim, S.; Inglett, G. E. Molecular weight and ionic strength dependence of fluorescence intensity of the calcofluor/β-glucan complex in flow-injection analysis. *Journal of Food Composition and Analysis*, 2006, *19*, 466–472.

Kivelä, R.; Gates, F.; Sontag-Strohm, T. Degradation of cereal beta-glucan by ascorbic acid induced oxygen radicals. *Journal of Cereal Science*, 2009a, *49*, 1–3.

Kivelä, R.; Nyström, L.; Salovaara, H.; Sontag-Strohm, T. Role of oxidative cleavage and acid hydrolysis of oat beta-glucan in modelled beverage conditions. *Journal of Cereal Science*, 2009b, *50*, 190–197.

Kivelä, R.; Pitkänen, L.; Laine, P.; Aseyev, V.; Sontag-Strohm, T. Influence of homogenisation on the solution properties of oat *β*-glucan. *Food Hydrocolloids*, 2010, *24*, 611–618.

Knuckles, B. E.; Chiu, M. M.; Betschart, A. A. *β*-Glucan enriched fractions from laboratory-scale dry milling and sieving of barley and oats. *Cereal Chemistry*, 1992, *69*, 198–202.

Kontogiorgos, V.; Biliaderis, C. G.; Kiosseoglou, V.; Doxastakis, G. Stability and rheology of egg-yolk-stabilized concentrated emulsions containing cereal *β*-glucans of varying molecular size. *Food Hydrocolloids*, 2004, *18*, 987–998.

Kontogiorgos, V.; Tosh, S. M.; Wood, P. J. Phase behaviour of high molecular weight oat *β*-glucan/whey protein isolate binary mixtures. *Food Hydrocolloids*, 2009a, *23*, 949–956.

Kontogiorgos, V.; Tosh, S. M.; Wood, P. J. Kinetics of phase separation of oat *β*-glucan/whey protein isolate binary mixtures. *Food Biophysics*, 2009b, *4*, 240–247.

Lambo, A. M.; Öste, R.; Nyman, M. E. Dietary fibre in fermented oat and barley *β*-glucan rich concentrates. *Food Chemistry*, 2005, *89*, 283–293.

Lazaridou, A.; Biliaderis, C. G. Cryogelation of cereal b-glucans: structure and molecular size effects. *Food Hydrocolloids*, 2004, *18*, 933–947.

Lazaridou, A.; Biliaderis, C. G. Molecular aspects of cereal *β*-glucan functionality: Physical properties, technological applications and physiological effects. *Journal of Cereal Science*, 2007a, 46, 101–118.

Lazaridou, A.; Biliaderis, C. G. Cryogelation phenomena in mixed skim milk powder – barley *β*-glucan – polyol aqueous dispersions. *Food Research International*, 2007b, *40*, 793–802.

Lazaridou, A.; Biliaderis, C. G. Concurrent phase separation and gelation in mixed oat *β*-glucans/sodium caseinate and oat *β*-glucans/pullulan aqueous dispersions. *Food Hydrocolloids*, 2009, *23*, 886–895.

Lazaridou, A.; Biliaderis, C. G.; Izydorczyk, M. S. Molecular size effects on rheological properties of oat *β*-glucans in solution and gels. *Food Hydrocolloids*, 2003, *17*, 693–712.

Lazaridou, A.; Biliaderis, C. G.; Micha-Screttas, M; Steele, B. R. A comparative study on structure-function relations of mixed-linkage (1→3),(1→4) linear *β*-D-glucans. *Food Hydrocolloids*, 2004, *18*, 837–855.

Lazaridou, A.; Duta, D.; Papageorgiou, M.; Belc, N.; Biliaderis, C. G. Effects of hydrocolloids on dough rheology and bread quality parameters in gluten-free formulations. *Journal of Food Engineering*, 2007, *79*, 1033–1047.

Lazaridou, A.; Vaikousi, H.; Biliaderis, C. G. Effects of polyols on cryostructurization of barley *β*-glucans. *Food Hydrocolloids*, 2008, *22*, 263–277.

Lee, S.; Kim, S.; Inglett, G. E. Effect of shortening replacement with oatrimon the physical and rheological properties of cakes. *Cereal Chemistry*, 2005, *82*, 120–124.

Lee, S.; Inglett, G. E. Effect of an oat *β*-glucan-rich hydrocolloid (C-trim30) on the rheology and oil uptake of frying batters. *Journal of Food Science*, 2007, *72*, E222–E226.

Lee, S.; Biresaw, G.; Kinney, M. P.; Inglett, G. E. Effect of cocoa butter replacement with a *β*-glucan-rich hydrocolloid (C-trim30) on the rheological and tribological properties of chocolates. *Journal of the Science of Food and Agriculture*, 2009a, *89*, 163–167.

Lee, S.; Inglett, G. E.; Palmquist, D.; Warner, K. Flavor and texture attributes of foods containing *β*-glucan-rich hydrocolloids from oats. *LWT - Food Science and Technology*, 2009b, *42*, 350–357.

Li, W.; Wang, Q.; Cui, S. W.; Huang, X.; Kakuda, Y. Elimination of aggregates of (1→3) (1→4)-*β*-D-glucan in dilute solutions for light scattering and size exclusion chromatography study. *Food Hydrocolloids*, 2006a, *20*, 361–368.

Li, W.; Cui, S. W.; Wang, Q. Solution and Conformational Properties of Wheat *β*-D-glucans Studied by Light Scattering and Viscometry. *Biomacromolecules*, 2006b, *7*, 446–452.

Li, W.; Wang, Q.; Cui, S. W.; Burchard, W.; Yada, R. Carbanilation of cereal *β*-glucans for molecular weight determination and conformational studies. *Carbohydrate Research*, 2007, *342*, 1434–1441.

Li, W; Cui, S. W.; Wang, Q.; Yada, R. Y. Studies of aggregation behaviours of cereal *β*-glucans in dilute aqueous solutions by light scattering: Part I. Structure effects. *Food Hydrocolloids*, 2011, *25*, 189–195.

Lozinsky, V. I.; Damshkaln, L. G.; Brown, R.; Norton, I. T. Study of cryostructuration of polymer systems. XXI. Cryotropic gel formation of the water–maltodextrin systems. *Journal of Applied Polymer Science*, 2002, *83*, 1658–1667.

Mårtensson, O.; Öste, R.; Holst, O. The effect of yoghurt culture on the survival of probiotic bacteria in oat-based, non-dairy products. *Food Research International*, 2002, *35*, 775–784.

McCleary, B. V.; Codd, R. Measurement of (1→3),(1→4)-*β*-D-glucan in barley and oats: A streamlined enzymic procedure. *Journal of the Science of Food and Agriculture*, 1991, *55*, 303–312.

Miyoshi, E.; Nishinari, K. Non-Newtonian flow behaviour of gellan gum aqueous solutions. *Colloid and Polymer Science*, 1999, *277*, 727–734.

Morgan, K. R., Ofman, D. J. Glucage, A Gelling *β*-glucan from Barley. *Cereal Chemistry*, 1998, *75*, 879–881.

Morgan, K. R., Roberts, C. J., Tendler, S. J. B.; Davies, M. C.; Williams, P.M. A ^{13}C CP/MAS NMR spectroscopy and AFM study of the structure of Glucagel^{TM}, a gelling *β*-glucan from barley. *Carbohydrate Research*, 1999, *315*, 169–179.

Morris, E. R.; Cutler, A. N.; Ross-Murphy, S. B.; Rees, D. A. Concentration and shear rate dependence of viscosity in random coil polysaccharide solutions. *Carbohydrate Polymers*, 1981, *1*, 5–21.

Murphy, E. A.; Davis, J. M.; Brown, A. S.; Carmichael, M. D.; Mayer, E. P.; Ghaffar, A. Effects of moderate exercise and oat *β*-glucan on lung tumor metastases and macrophage antitumor cytotoxicity. *Journal of Applied Physiology*, 2004, *97*, 955–959.

Naumann, E.; van Rees, A. B.; Önning, G.; Oste, R.; Wydra, M.; Mensink, R. P. *β*-Glucan incorporated into a fruit drink effectively lowers serum LDL-cholesterol concentrations. *American Journal of Clinical Nutriton*, 2006, *83*, 601–605.

Panahi, S.; Ezatagha, A.; Temelli, F.; Vasanthan, T.; Vuksan, V. *β*-Glucan from two sources of oat concentrates affect postprandial glycemia in relation to the level of viscosity. *Journal of the American College of Nutrition*, 2007, *26*, 639–644.

Papageorgiou, M.; Lakhdara, N.; Lazaridou, A.; Biliaderis, C. G.; Izydorczyk, M. S. Water extractable (1→3,1→4)-*β*-D-glucans from barley and oats: An intervarietal study on their structural features and rheological behaviour. *Journal of Cereal Science*, 2005, *42*, 213–224.

Park, S. Y.; Bae, I. Y.; Lee, S. H.; Lee. G. Physicochemical and hypocholesterolemic characterization of oxidized oat *β*-glucan. *Journal of Agricultural and Food Chemistry*. 2009, *57*, 439–443.

Peterson, D. M. Malting oats: effects on chemical composition of hull-less and hulled genotypes. *Cereal Chemistry*, 1998, *75*, 230–234.

Pillai, R.; Redmond, M.; Röding, J. Anti-wrinkle therapy: significant new findings in the non-invasive cosmetic treatment of skin wrinkles with beta-glucan. *International Journal of Cosmetic Science*, 2005, *27*, 292.

Piñero, M. P.; Parra, K.; Huerta-Leidenz, N.; Arenas de Moreno, L.; Ferrer, M.; Araujo, S.; Barboza, Y. Effect of oat's soluble fibre (*β*-glucan) as a fat replacer on physical, chemical, microbiological and sensory properties of low-fat beef patties. *Meat Science*, 2008, *80*, 675–680.

Pizzoferrato, L.; Manzi, P.; Bertocchi, F.; Fanelli, C.; Rotilio, G.; Paci, M. Solid-state 13C CP MAS NMR spectroscopy of mushrooms gives directly the ratio between proteins and polysaccharides. *Journal of Agricultural and Food Chemistry*, 2000, *48*, 5484–5488.

Queenan, K. M.; Stewart, M. L.; Smith, K. N.; Thomas, W.; Fulcher, R. G.; Slavin, J. L. Concentrated oat *β*-glucan, a fermentable fiber, lowers serumcholesterol in hypercholesterolemic adults in a randomized controlled trial. *Nutrition Journal*, 2007, *6*, 6–13.

Ramakers, J. D.; Volman, J. J.; Önning, G.; Biörklund, M.; Mensink, R. P.; Plat, J. Immune-stimulating effects of oat *β*-glucan on enterocytes. *Molecular Nutrition and Food Research*, 2007, *51*, 211–220.

Rimsten, L.; Haraldsson, A.; Andersson, R.; Alminger, M.; Sandberg, A.; Åman, P. Effects of malting on *β*-glucanase and phytase activity in barley grain. *Journal of the Sciene of Food and Agriculture*. 2002, *82*, 904–912.

Rimsten, L.; Stenberg, T.; Andersson, R.; Andersson, A.; Åman, P. Determination of *β*-glucan molecular weight using SEC with calcofluor detection in cereal extracts. *Cereal Chemistry*, 2003, *80*, 485–490.

Regand, A.; Tosh, S. M.; Wolever, T. M. S.; Wood, P. J. Physicochemical properties of β-glucan in differently processed oat foods influence glycemic response. *Journal of Agricultural and Food Chemistry*, 2009, *57*, 8831–8838.

Ren, Y.; Ellis, P. R.; Ross-Murphy, S. B.; Wang, Q.; Wood, P. J. Dilute and semi-dilute solution properties of (1→3),(1→4)-*β*-D-glucan, the endosperm cell wall polysaccharide of oats (*Avena sativa* L.). *Carbohydrate Polymers*, 2003, *53*, 401–408.

Ripsin, C. M.; Keenan, J. M.; Jacobs, D. R.; Elmer, P. J.; Welch, R. R.; Van Horn, L.; Liu, K.; Turnbull, W. H.; Thye, F. W.; Kestin, M.; Hegsted, M.;Davidson, D. M.; Davidson, M. H.; Dugan, L. D.; Demark-Wahnefried, W.; Beling, S. Oat products and lipid lowering: a meta-analysis. *Journal of American Medical Associstion*. 1992, *267*, 3317–3325.

Roubroeks, J. R.; Andersson, R.; Åman, P. Structural features of (1-3),(1-4)-β-D-glucan and arabinoxylan fractions isolated from rye bran. *Carbohydrate Polymers*, 2000a, *42*, 3–11.

Roubroeks, J. R.; Mastromauro, D. I.; Andersson, R.; Christensen, B. E.; Åman, P. Molecular weight, structure, and shape of oat (1→3),(1→4)-*β*-D-glucan fractions obtained by enzymatic degradation with lichenase. *Biomacromolecules*, 2000b, *1*, 584–591.

Roubroeks, J. P.; Skjåk-Braek, G.; Ryan, L.; Christensen, B. E. Molecular weight dependency on the production of the TNF stimulated by fractions of rye (1→3),(1→4)-*β*-D-glucan. *Scandinavian Journal of Immunology*, 2000c, *52*, 584–587.

Roubroeks, J. P.; Andersson, R.; Mastromauro, D. I.; Christensen, B. E.; Åman, P. Molecular weight, structure and shape of oat (1→3),(1→4)-*β*-D-glucan fractions obtained by enzymatic degradation with (1→4)-*β*-D-glucan 4-glucanohydrolase from *Trichoderma reesei*. *Carbohydrate Polymers*, 2001, *46*, 275–285.

Sayar, S.; Jannink, J.-L.; White, P. J. In vitro bile acid binding of flours from oat lines varying in percentage and molecular weight distribution of *β*-glucan. *Journal of Agricultural and Food Chemistry*, 2005, *53*, 8797–8803.

Sayar, S.; Jannink, J.-L.; White, P. J. In vitro bile acid binding activity within flour fractions from oat lines with typical and high *β*-glucan amounts. *Journal of Agricultural and Food Chemistry*, 2006, *54*, 5142–5148.

Sayar, S.; Jannink, J.-L.; White, P. J. Digestion residues of typical and high-β-glucan oat flours provide substrates for *in vitro* fermentation. *Journal of Agricultural and Food Chemistry*, 2007, *55*, 5306–5311.

Schmitt, M R.; Wise, M. L. Barley and oat *β*-glucan content measured by calcofluor fluorescence in a microplate assay. *Cereal Chemistry*, 2009, *86*, 187–190.

Shiroishi, M.; Amano, Y.; Hoshino, E.; Nisizawa, K.; Kanda, T. Hydrolysis of various celluloses, (1→3),(1→4)-*β*-D-glucans, and xyloglucan by three endo-type cellulases. *Mokuzai Gakkaishi*, 1997, *43*, 178–187.

Skendi, A.; Biliaderis, C. G.; Lazaridoub, A.; Izydorczyk, M. S. Structure and rheological properties of water soluble β-glucans from oat cultivars of *Avena sativa* and *Avena bysantina*. *Journal of Cereal Science*, 2003, *38*, 15–31.

Smith, K. N.; Queenan, K. M.; Thomas, W.; Fulcher, R. G.; Slavin, J. L. Physiological effects of concentrated barley β-glucan in mildly hypercholesterolemic adults. *Journal of the American College of Nutrition*, 2008, *27*, 434–440.

Snart, J.; Bibiloni, R.; Grayson, T.; Lay, C.; Zhang, H.; Allison, G. E.; Laverdiere, J. K.; Temelli, F.; Vasanthan, T.; Bell, R.; Tannock G. W. Supplementation of the diet with high-viscosity beta-glucan results in enrichment for lactobacilli in the rat cecum. *Applied and Environmental Microbiology*, 2006, *72*, 1925–1931.

Staudte, R. G.; Woodward, J. R.; Fincher, G. B.; Stone, B. A. Water soluble (1-3)(1-4)-β-D-glucans from barley (*Hordeum vulgare*) endosperm. III. Distribution of cellotriosyl and cellotetraosyl residues. *Carbohydrate Polymers*, 1983, *3*, 299–312.

Sørensen, I.; Pettolino, F. A.; Wilson, S. M.; Doblin, M. S.; Johansen, B.; Bacic, A.; Willats W. G. T. Mixed-linkage (1→3),(1→4)-β-D-glucan is not unique to the Poales and is an abundant component of *Equisetum arvense* cell walls. *The Plant Journal*, 2008, *54*, 510–521.

Tappy, L.; Gügolz, E.; Würsch, P. Effects of breakfast cereals containing various amounts of beta-glucan fibers on plasma glucose and insulin responses in NIDDM subjects. *Diabetes Care*, 1996, *19*, 831–834.

Tietyen, J. L.; Nevins, D. J.; Shoemaker, C. F.; Schneeman, B. O. Hypocholesterolemic potential of oat bran treated with an endo-β-D-glucanase from *Bacillus subfilis*. *Journal of Food Science*, 1995, *60*, 558-560.

Timmermans, J. W.; van Leeuwen, M. B.; Tournois, H.; de Wit, D.; Vliegenthart, J. F. G. Quantification analysis of the molecular weight distribution of inulin by means of anion exchange HPLC with pulsed amperometric detection. *Journal of Carbohydrate Chemistry*, 1994, *13*, 881–888.

Tosh, S. M.; Wood, P. J.; Wang, Q.; Weisz, J. Structural characteristics and rheological properties of partially hydrolyzed oat β-glucan: the effects of molecular weight and hydrolysis method. *Carbohydrate Polymers*, 2004a, *55*, 425–436.

Tosh, S. M.; Brummer, Y.; Wood, P. J.; Wang, Q.; Weisz, J. Evaluation of structure in the formation of gels by structurally diverse (1→3)(1→4)-*β*-D-glucans from four cereal and one lichen species. *Carbohydrate Polymers*, 2004b, *57*, 249–259.

Tosh, S. M.; Brummer, Y.; Miller, S. S.; Regand, A.; Defelice, C.; Duss, R.; Wolever, T. M. S.; Wood, P. J. Processing affects the physicochemical properties of *β*-glucan in oat bran cereal. *Journal of Agricultural and Food Chemistry*, 2010, *58*, 7723–7730.

Tudorica, C. M.; Jones, T. E. R.; Kuri, V.; Brennan, C. S. The effects of refined barley *β*-glucan on the physico-structural properties of low-fat dairy products: curd yield, microstructure, texture and rheology. *Journal of the Science of Food and Agriculture*, 2004, *84*, 1159–1169.

Tvaroska, I.; Ogawa, K.; Deslandes, Y.; Marchessault, R. H. Crystalline conformation and structure of lichenan and barley *β*-glucan. *Canadian Journal of Chemistry*, 1983, *61*, 1608–1616.

Vachon, C.; Jones, J. D.; Wood, P. J.; Savoie, Laurent. Concentration effect of soluble dietary fibers on postprandial glucose and insulin in the rat. *Canadian Journal of Physiology and Pharmacology*, 1988, *66*, 801–806.

Vaikousi, H.; Biliaderis, C. G. Processing and formulation effects on rheological behavior of barley *β*-glucan aqueous dispersions. *Food Chemistry*, 2005, *91*, 505–516..

Vaikousi, H.; Biliaderis, C. G.; Izydorczyk, M. S. Solution flow behavior and gelling properties of water-soluble barley (1→3,1→4)-*β*-glucans varying in molecular size. *Journal of Cereal Science*, 2004, *39*, 119–137.

Vårum, K. M.; Martinsen, A.; Smidsrød, O. Fractionation and viscometric characterization of a (1→3), (1→4)-*β*-D-glucan from oat, and universal calibration of a high-performance size-exclusion chromatographic system by the use of fractionated *β*-glucans, alginates and pullulans. *Food Hydrocolloids,* 1991, *5*, 363–374.

Vårum, K. M.; Smidsrød, O.; Brant, D. A. Light scattering reveals micellelike aggregation in the (1→3), (1→4)-*β*-D-glucans from oat aleurone. *Food Hydrocolloids*, 1992, *5*, 497–511.

Vasanthan, T.; Temelli, F. Grain fractionation technologies for cereal beta-glucan concentration. *Food Research International*, 2008, *41*, 876–881.

Virkkia ,L.; Johanssona, L.; Ylinena, M.; Maunub, S.; Ekholm, P. Structural characterization of water-insoluble nonstarchy polysaccharides of oats and barley. *Carbohydrate Polymers*, 2005, *59*, 357–366.

Volikakis, P.; Biliaderis, C. G.; Vamvakas, C.; Zerfiridis, G. K. Effects of a commercial oat-*β*-glucan concentrate on the chemical, physico-chemical and sensory attributes of a low-fat white-brined cheese product. *Food Research International*, 2004, *37*, 83–94.

Volman, J. J.; Ramakers, J. D.; Plat, J. Dietary modulation of immune function by β-glucans. *Physiology and Behavior*, 2008, *94*, 276–284.

Volman, J. J.; Mensink, R. P.; Buurman, W. A.; Onning, G.; Plat, J. The absence of functional dectin-1 on enterocytes may serve to prevent intestinal damage. *European Journal of Gastroenterology and Hepatology*, 2010a, *22*, 88–94.

Volman, J. J.; Mensink, R. P.; Ramakers, J. D.; de Winther, M. P.; Carlsen, H.; Blomhoff, R.; Buurman, W.A.; Plat, J. Dietary (1→3), (1→4)-*β*-D-glucans from oat activate nuclear factor-κB in intestinal leukocytes and enterocytes from mice. *Nutrition Research*, 2010b, *30*, 40–48.

Wang, L.; Newman, R. K.; Newman, C. W.; Hofer, P. J. Barley *β*-glucans alter intestinal viscosity and reduce plasma cholesterol concentrations in chicks. *Journal of Nutrition*, 1992, *122*, 2292–2297.

Wang, Q.; Wood, P. J.; Cui, W. Microwave assisted dissolution of *β*-glucan in water — implications for the characterisation of this polymer. *Carbohydrate Polymers*, 2002, *47*, 35–38.

Wang, Q.; Wood, P. J.; Huang, X.; Cui, W. Preparation and characterization of molecular weight standards of low polydispersity from oat and barley (1→3)(1→4)-*β*-D-glucan. *Food Hydrocolloids*, 2003, *17*, 845–853.

Wang, R.; Koutinas, A. A.; Campbell, G. M. Dry processing of oats – Application of dry milling. *Journal of Food Engineering*, 2007, *82*, 559–567.

Wang, S. C.; Bligh, S. W.; Zhu, C. L.; Shi, S. S.; Wang, Z.T.; Hu, Z. B.; Crowder, J.; Branford-White, C.; Vella, C. Sulfated *β*-glucan derived from oat bran with potent anti-HIV activity. *Journal of Agricultural and Food Chemistry*, 2008, *56*, 2624–2629.

Wang, H.-C.; Hung, C.-H.; Hsu, J.-D.; Yang, M.-Y.; Wang, S.-J.; Wang, C.-J. Inhibitory effect of whole oat on aberrant crypt foci formation and colon tumor growth in ICR and BALB/c mice. *Journal of Cereal Science*, 2010, doi:10.1016/j.jcs.2010.09.009

Warner, K. Inglett, G. E. Flavor and texture characteristics of foods containing Z-trim corn and oat fibers as fat and flour replacers. *Cereal Foods World*, 1997, *42*, 821–825.

Westerlund, E.; Andersson, R.; Åman, P. Isolation and chemical characterization of water-soluble mixed-linked *β*-glucans and arabinoxylans in oat milling fractions. *Carbohydrate Polymers*, 1993, *20*, 115–123.

Wilhelmson, A.; Oksman-Caldentey, K. M.; Laitila, A.; Suortti, T.; Kaukovirta-Norja, A.; Poutanen, K. Development of a germination process for producing high β-glucan, whole grain food ingredients from oat. *Cereal Chemistry*, 2001, *78*, 715–720.

Wilson, T. A.; Nicolosi, R. J.; Delaney, B.; Chadwell, K.; Moolchandani, V.; Kotyla, T.; Ponduru, S.; Zheng, G.-H.; Hess, R.; Knutson, N.; Curry, L.; Kolberg, L.; Goulson, M.; Ostergren, K. Reduced and high molecular weight barley β-glucans decrease plasma total and non-HDL-Cholesterol in hypercholesterolemic syrian golden hamsters. *Journal of Nutrition*, 2004, *134*, 2617–2622.

Wolever, T. M. S.; Spadafora, P.; Eshuis, H. Interaction between colonic acetate and propionate in humans. *American Journal of Clinical Nutrition*, 1991, *53*, 681–687.

Wolever, T. M. S; Tosh, S. M.; Gibbs, A. L.; Brand-Miller, J.; Duncan, A. M.; Hart, V.; Lamarche, B.; Thomson, B. A.; Duss, R.; Wood P. J. Physicochemical properties of oat β-glucan influence its ability to reduce serum LDL cholesterol in humans: a randomized clinical trial. *American Journal of Clinical Nutrition*, 2010, *92*, 723–732.

Wood, P. J. Relationships between solution properties of cereal β-glucans and physiological effects – a review. *Trends in Food Science and Technology*, 2004, *15*, 313–320.

Wood, P. J.; Siddiqui, I. R.; Paton, D. Extraction of high-viscosity gums from oats. *Cereal Chemistry*, 1978, *55*, 1038–1049.

Wood, P. J.; Fulcher, R. G.; Stone, B. A. Studies on the specificity of interaction of cereal cell wall components with Congo Red and Calcoflour: specific detection and histochemistry of (1→3)(1→4)-β-D-glucan. *Journal of Cereal Science*, 1983, 1, 95–110.

Wood, P. J.; Weisz, J.; Fedec, P.; Burrows, V. D. Large-Scale preparation and properties of oat fractions enriched in (1→3)(1→4)-β-D-glucan. *Cereal Chemistry*, 1989, *66*, 97–103.

Wood, P. J.; Weisz, J.; Blackwell, B. A. Molecular characterization of cereal β-glucans. Structural analysis of oat β-D-glucan and rapid structural evaluation of β-D-glucans from different sources by high performance liquid chromatography of oligosaccharides released by lichenase. *Cereal Chemistry*, 1991a, *68*, 31–39.

Wood, P. J.; Weisz, J.; Mahn, W.. Molecular characterisation of cereal β-glucans. II. Size-exclusion chromatography for comparison of molecular weight. *Cereal Chemistry*, 1991b, *68*, 530–536.

Wood, P. J.; Weisz, J.; Blackwell, B. A. Structural studies of (1→3) (1→4)-β-D-glucans by ^{13}C-nuclear magnetic resonance spectroscopy and by rapid analysis of cellulose-like regions using high-performance anion-exchange chromatography of oligosaccharides released by lichenase. *Cereal Chemistry*, 1994a, *71*, 301–307.

Wood, P. J.; Braaten, J. T.; Scott, F. W.; Riedel, K. D.; Wolynetz, M. S.; Collins, M. W. Effect of dose and modification of viscous properties of oat gum on plasma glucose and insulin following an oral glucose load. *British Journal of Nutrition*, 1994b, *12*, 731–743.

Wood, P. J.; Beer, M. U.; Butler, G. Evaluation of role of concentration and molecular weight of oat β-glucan in determining effect of viscosity on plasma glucose and insulin following an oral glucose load. *British Journal of Nutriton*, 2000, *84*, 19–23.

Woodward, J. R.; Fincher, G. B.; Stone, B. A. Water soluble (1→3)(1→4)-β-D-glucans from barley (*Hordeum vulgare*) endosperm. II. Fine structure. *Carbohydrate Polymers*, 1983, *3*, 207–25.

Wright, R. S.; Anderson, J. W.; Bridges, S. R. Propionate inhibits hepatocyte lipid synthesis. *Proceedings of the Society for Experimental Biology and Medicine*, 1990, *195*, 26–29.

Wu, Y. V.; Doehlert, D. C. Enrichment of β-glucan in oat bran by fine grinding and air classification. *Lebensmittel-Wissenschaft und-Technologie*, 2002, *35*, 30–33.

Wu, J.; Zhang, Y.; Wang, L.; Xie, B.; Wang, H.; Deng, S. Visualization of single and aggregated hulless oat (*Avena nuda* L.) (1→3),(1→4)-β-D-glucan molecules by atomic force microscopy and confocal scanning laser microscopy. *Journal of Agricultural and Food Chemistry*, 2006, *54*, 925–934.

Wu, J.; Deng, X.; Tian, B.; Wang, L.; Xie, B. Interactions between oat β-glucan and calcofluor characterized by spectroscopic method. *Journal of Agricultural and Food Chemistry*, 2008, *56*, 1131–1137.

Wu, J.; Deng, X.; Zhang, Y.; Wang L.; Tian B.; Xie B. Application of Atomic Force Microscopy in the Study of Polysaccharide. *Agricultural Sciences in China*, 2009, *8*, 1458–1465.

Yokoyama, W. H.; Hudson, C. A.; Knuckles, B. E.; Chiu, M.-C. M.; Sayre, R. N.; Turnlund, J. R.; Schneeman, B. O. Effect of Barley beta-glucan in Durum Wheat Pasta on Human Glycemic Response. *Cereal Chemistry*, 1997, *74*, 293–296.

Yokoyama, W. H.; Knuckles, B. E.; Stafford, A.; Inglett, G. Raw and processed oat ingredients lower plasma cholesterol in the hamster. *Journal of Food Science*, 1998, *63*, 713–715.

Yoo, D.; Lee, B.; Chang, P.; Lee, H. G.; Yoo, S. Improved quantitative analysis of oligosaccharides from lichenase-hydrolyzed water-soluble barley β-glucans by high-performance anion-exchange chromatography. *Journal of Agricultural and Food Chemistry*, 2007, *55*, 1656–1662.

Yun, C.-H.; Estrada, A.; Van Kessel, A.; Gajadhar, A. A.; Redmond, M. J.; Laarveld, B. β-(1→3, 1→4) oat glucan enhances resistance to *Eimeria vermiformis* infection in immunosuppressed mice. *International Journal for Parasitology*, 1997, *27*, 329–337.

Yun, C.-H.; Estrada, A.; Van Kessel, A.; Park, B.-C.; Laarveld, B. β-Glucan, extracted from oat, enhances disease resistance against bacterial and parasitic infections. *FEMS Immunology and Medical Microbiology*, 2003, *35*, 67–75.

Zhang, D.; Doehlert, D. C.; Moore, W. R. Rheological properties of (1→3),(1→4) -β-D-glucans from raw, roasted and steamed oat groats. *Cereal Chemistry*, 1998, 75, 433–438.

In: Oats: Cultivation, Uses and Health Effects ISBN 978-1-61324-277-3
Editor: D. L. Murphy, pp. 51-96

Chapter 2

STABILITY AND DEGRADATION OF SOLUBLE β-GLUCAN IN AQUEOUS PROCESSING

R. Kivelä* and T. Sontag-Strohm
Department of Food and Environmental Sciences,
Group of Cereal Technology,
University of Helsinki, Finland

ABSTRACT

The most abundant soluble dietary fibre of oat, (1→4),(1→3)-β-D-glucan, is a linear polysaccharide, which is evidently beneficial for human health. The health promoting effects are generally related to ability of beta-glucan to form highly viscous solutions, which in many foods define technological functionality of beta-glucan as well. The solution properties are influenced by parameters such as molar mass, extractability, solubility and structure of beta-glucan. Degradation of beta-glucan influences molar mass, but also its extractability and conformation in certain conditions. Food manufacturing processes cause degradation, which may be a threat for the functionality of beta-glucan if the degradation mechanisms were not understood. Degradation may as well serve tools for modification and innovation as soon as it is well-managed. However, only few degradation mechanisms are generally

*Correspondence:Tel.:+358-9-19158540, Fax:+358-9-19158460, Email: reetta.kivela@helsinki.fi

considered, and non-enzymatic degradation such as oxidation has been highly neglected in beta-glucan related literature until our recent publications. This review will discuss process-induced degradation of oat beta-glucan concentrating on aqueous processing. Biodegradation, chemical degradation, thermal degradation, mechanical energy induced degradation and oxidative cleavage will be discussed as degradation mechanisms of oat beta-glucan.

1. Introduction

β-Glucan typically forms 3,5 - 5,0 % of the dry matter content of whole oat grain. The β-glucan content is affected by several factors such as variation between cultivar and the growing conditions (Asp, Mattsson, & Önning, 1992). B-glucan is distributed throughout the starchy endosperm in the cell walls, but is concentrated in the bran (aleurone and sub-aleurone layer) fractions of the oat grain (Figure 1).

Oat bran fractions roughly consist of 6-8% of β glucan, 15-20% oat bran concentrates and up to 80% β-glucan isolates (Lazaridou, Biliaderis, & Izydorczyk, 2007). Interest in developing β-glucan rich fractions and further β-glucan enriched products originates from the evidenced health benefit of oat. Several authorities have proclaimed the serum cholesterol lowering or maintenance effect of β-glucan to be scientifically well evidenced [(FDA, 1997); (SNF, 2001); (JHCI, 2006); (Voedingscentrum, 2005); (SFOPH, 2006); (AFSSA, 2008); (EFSA, 2009)].

The serum glucose attenuating ability of β-glucan has also been widely studied, and positively recognized by the Swedish Code of Practice (SkåneDairy, 2002). In addition to the cholesterol and postprandial glucose lowering effects, cereal β-glucan has been shown to reduction of blood pressure of humans (Keenan, Pins, Frazel, Moran, & Turnquist, 2002), increased satiety of humans (Beck;Tosh;Batterham;Tapsell;& Huang, 2009), control body weight of mice (Bae, Lee, Kim, & Lee, 2009), and anti-inflammatory effects in mice (Davis, Murphy, Brown, Carmichael, Ghaffar, & Mayer, 2004).

The effective daily dose of β-glucan in relation to its cholesterol lowering ability is generally accepted to be 3 g (Ripsin, et al., 1992). However, since the solution properties of β-glucan are evidently altered by processing, but the mechanisms are not fully understood, the effectiveness of oat β-glucan is generally limited to oat bran and other "minimally processed products" as worded by the (EFSA, 2009). This review introduces the β-glucan molecule and its changes during processing and discusses the degradation mechanisms relevant to β-glucan in aqueous environment.

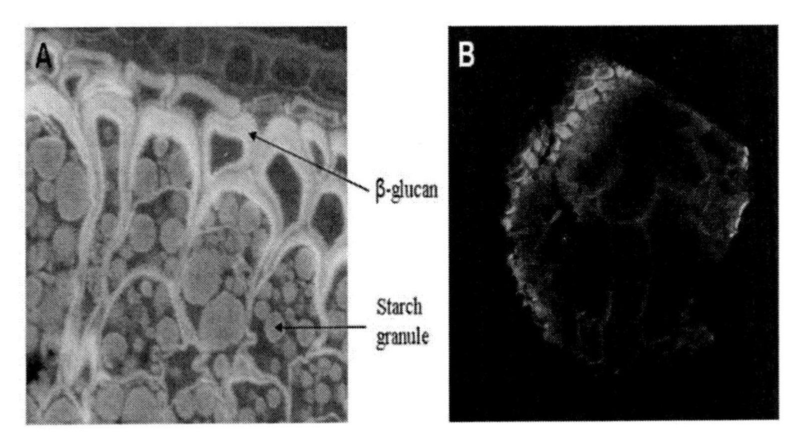

Figure 1. A) Distribution of beta-glucan in the cell walls and bran layers of an oat kernel (adapted from www.oatsandhealth.org by courtesy of VTT Technical Research Centre of Finland). B) Confocal scanning laser microscopy image of an oat bran particle from porridge showing the beta-glucan in the cell walls in blue. The image size is 85 μm x 85 μm (unpublished material of Kivelä/University of Helsinki).

2. Oat B-Glucan in Aqueous Solutions

2.1. Molecular Properties of Oat β-Glucan

Oat β-glucan is a linear polysaccharide that consists of β-D-glucopyranosyl units. The units are joined by either (1→3)- or (1→4)-β-D-linkages, hence the name mixed-linkage (1→3,1→4)-β-D-β-glucan. The (1→4)-β-linkages dominate and mostly occur in groups of two or three, forming segments with three and four glucose residues (DP3/cellotrioses and DP4/cellotetraoses). These cellulose-like sequences are interrupted by separated (1→3)- β-linkages as first demonstrated by (Parrish, Perlin, & Reese, 1960) and later by others. The (1→3)- β-linkages increase the flexibility of the chain preventing the close packing known for cellulose molecules (Buliga, Brant, & Fincher, 1986), and enabling the water-solubility of cereal β-glucan. The ratio of (1→4)-β-linkages to (1→3)- β-linkages has been demonstrated to be approximately 70:30, and the main building blocks cellotrioses and cellotetraoses comprise over 90% of the molecule. A minor part of the oat β-glucan structure consists of longer cellulose-like sequences up to DP20, of which DP5, DP6 and DP9 are the most abundant the ratios varying between cultivars ((Doublier & Wood, 1995); (Lazaridou, Biliaderis, & Izydorczyk, 2003), (Fincher, 2009)). The weight average molar mass of oat β-glucan from oat is typically approximately 1000-3000 x 10^3 g/mol. For example, the molar masses of β-glucan in extracts of oat whole flour, rolled oats, oat bran and oat bran concentrates were found to be between 2060-2300 x 10^3g/mol, determined as relative weight average molar masses (HPSEC-fluoresence) after aqueous extraction by amylases and proteases (Åman, Rimsten, & Andersson, 2004). Wood (1991) reported molar masses of 2900-3100 x 10^3g/mol from oat groat and bran after alkaline extraction of various cultivars. Unlike the grain fraction or the cultivar, the extraction method significantly influences the magnitude of the molar mass of β-glucan. (Wood P. , 1991) observed that ethanol precipitation decreased the molar mass from ≈3000 x10^3g/mol to 2100x10^3g/mol and the $(NH_4)_2SO_4$ -purification step to 400 x10^3g/mol.

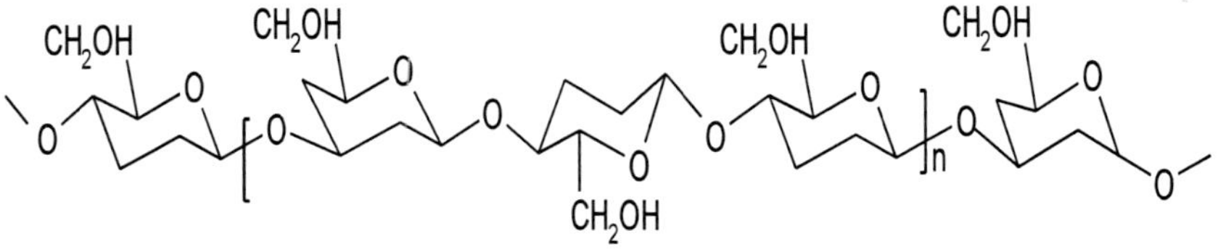

Scheme 1. Structure of the (1→3,1→4)-β-D-glucan.

In water solutions, β-glucan occurs as a single-stranded, partially stiff worm-like molecule and an expanded coil having Mark-Houwink's exponent of 0.6-0.7 in the order oat > barley > wheat (Table 1, (Gómez, Navarro, Manzanares, Horta, & Carbonell, 1997b); (Böhm & Kulicke, 1999a); (Wu, Zhang, Wang, Xie, Wang, & Deng, 2006), (Li, Cui, Wang, & Yada, 2010)). The exponent α in the Mark-Houwink equation $[\eta] = KM_w^{\alpha}$, which relates the intrinsic viscosity (i.e. the volume which is occupied by a molecule in solution) and molar mass, is in the range of $0.5 \leq \alpha \leq 0.8$ for flexible molecules and random coils. Values of 0.70–0.75 have also been widely reported for other linear polysaccharides such as galactomannan (Picout & Ross-Murphy, 2007). The expansion is likely due to the partial stiffness, which may be caused by the rigid structures of the 1,4-β-linked segments. When the conformation of oat β-glucan was examined in SDS solution in order to investigate the characteristics of monomers, a rigid rodlike chain was found, probably due to the elimination of intra-actions by the alkaline conditions (Wu, Zhang, Wang, Xie, Wang, & Deng, 2006). Properties of the molecule can be determined also by a shape parameter $\rho=R_g/R_h$, which depends on the chain architechture, conformation and polydispersity, but not on the molar mass (Burchard, 1995). For polydisperse random coils, $\rho=2.05$ in a good solvent and $\rho=1.73$ in a theta-solvent (where polymer coils act like ideal chains), while for star-branched structures $\rho \approx 1$ and for a rigid sphere $\rho \approx 0.7$. Moreover, ρ-values of 1.5-1.7 have been determined for oat β-glucan, $\rho=1.7$ for wheat unimer, $\rho \approx 2.2$ for barley β-glucan unimer and $\rho \approx 0.9$ for aggregates ((Grimm, Krüger, & Burchard, 1995); (Li, Wang, Cui, Huang, & Kakuda, 2006); (Kivelä, Pitkänen, Laine, Aseyev, & Sontag-Strohm, 2010); Table 1).

2.2. Aggregation and Gelation

The challenge in determining the molecular properties of β-glucan in water solution is not only its partial insolubility, but also its tendency to aggregate. In dilute aqueous solutions ($c<c^*$), oat and barley β-glucans have been reported to occur as fringed micelle-type aggregates, which grow by side-to-side via hydrogen bonding of the cellotriose-sequences (Figure 2; (Grimm, Krüger, & Burchard, 1995); (Böhm & Kulicke, 1999a); (Wu, Zhang, Wang, Xie, Wang, & Deng, 2006). (Vårum, Smidsrod, & Brant, 1992)(1992) observed that only a fraction of the molecules was involved in association to form large stable aggregates, since light scattering detected aggregates that were not revealed by osmotic pressure measurements. The conclusions highlighted the problems of molecular association from the perspective of analytics, since the intensity of scattered light is proportional to the weight average molar mass, and thus even a very small number of aggregates may lead to inaccurate results in molecular weight determination. For this reason solvents such as cadoxen, cuoxam and 0.5 M NaOH have been used in studies of aggregation by light scattering without a fractionation method ((Grimm, Krüger, & Burchard, 1995); (Li, Wang, Cui, Huang, & Kakuda, 2006); (Li, Cui, Wang, & Yada, 2010). (Li, Cui, Wang, & Yada, 2010)concluded that the aggregation process was rapid in water, and the size of the aggregates was limited and concentration dependent due to the equilibrium between aggregate dissociation and association.

Table 1. Molecular parameters of oat β-glucan in aqueous solutions. Selected molar masses (M_w or M_p), the Mark Houwink exponent α, intrinsic viscosity *[η]*, value of coil overlap parameter *[η]*c in critical concentration c* and c, aggregation degree x ($size_{,unimer}$/$size_{, aggregate\ in\ water}$), the ratio of cellotrioses to tetraoses (DP3/DP4) are presented**

Mw (10^3g/mol)	α	[η] (dl/g)	DP3/DP4	c*[η] (g/dl)	c**[η] (g/dl)	x	Extract solvent	Analysis method	Reference
1160[a]			2.2				pH10	SEC/FI	Cui et al., 2000
1190		8.6	2.4				pH10	SEC/triple	Tosh et al.,2004
780		5.6	2.0	1.2	7.8		aq	SEC/MALLS/RI	Skendi et al., 2003
1200			2.4		2-4		pH10	SEC/LS/FI	Ren et al., 2003[b]
1200		9.6	2.4	0.7	2.5		pH10	SEC/LS/FI	Douplier& Wood, 1995
250		3.8	2.0	0.8	2.7		enz/aq	SEC/RI	Lazaridou et al., 2003
1200	0.57	11.0					pH10	SEC/LS/FI	Wang et al., 2001
456	0.71[c]	4.6					comm.	SEC/MALLS/RI	Gómez et al., 1997b
2025	0.58	4.2				4.2	$(NH_4)_2SO_4$	SLS/DLS	Li et al., 2011a
5400 (barley)		2.2				30.5	comm.	SLS/DLS	Grimm et al.,1995

[a]Mp (peak molar mass instead of weight average molar mass).
[b] Same material as studied in (Doublier & Wood, 1995).
[c] The determination mainly included barley beta-glucan samples, one oat beta-glucan was included.

Abbreviations: SEC=HPSEC, high performance size exclusion chromatography Fl=fluorescence detection, triple=RI-viscostec and dual angle light scattering detection SLS/DLS=static light scattering and dynamic light scattering studies pH10= extraction with sodium carbonate buffer enz.= extraction in buffers with digestion enzymes (NH4)2SO4 = ammonium sulphate precipitation after extraction to eliminate of arabinoxylans.

They found the degree of aggregation ($x = R_{h,\ monomer}/R_{h,\ aggregate\ in\ water}$, monomers in cadoxen solution) to be lower for the more rigid β-glucans in the order wheat (3.9) > barley (4.4) > oat (7.5), and thus concluded that the aggregation was controlled by the diffusion rate in dilute solutions. An aggregation degree ($x = M_{w,\ monomer}/M_{w,\ aggregate\ in\ water}$) of 17-70 was recorded for barley β-glucan when the monomer was determined in cuoxam (Grimm, Krüger, & Burchard, 1995) and of 2-4 for oat β-glucan as the monomer was determined in LiI-solution (Grimm, Krüger, & Burchard, 1995); Table 1). By others, the aggregation degree of barley β-glucan was found to be negligible and beta-glucan was reported to occur as non-associated in dilute solutions (Böhm & Kulicke, 1999a). The discrepancies in the results may partially be based on the note that aggregation degree up to $x=5$ may not change the chain flexibility, and may thus not be recognized by Mark-Houwink's exponent (Grimm, Krüger, & Burchard, 1995). However, Li, et al. (2006) demonstrated the degree of aggregation to be negligible in the conditions of high-performance size-exclusion chromatography (HPSEC), when compared to non-aggregates in NaOH measured by light scattering. The dissociating effect of the HPSEC-conditions, which combines shear flow, temperature and dilution before light scattering detection, was observed also by others ((Vårum, Smidsrod, & Brant, 1992); (Gómez, Navarro, Manzanares, Horta, & Carbonell, 1997a), (Kivelä, Sontag-Strohm, Loponen, Tuomainen, & Nyström, 2011a)).

The ratio of cellotrioses and cellotetraoses is a characteristic structural indicator of cereal β-glucans, which follows the order of wheat (4.2–4.5)< barley (2.8–3.3) < oat (2.0–2.4, Table 1). This ratio is related to solubility, which follows in the order wheat < barley < oat (Cui, Wood, Blackwell, & Nikiforuk, 2000)and gelling ability ((Böhm & Kulicke, 1999a)(Böhm & Kulicke, 1999b); (Tosh, Wood, Wang, & Weisz, 2004)). Notably, solubility has also been suggested to be dependent also on other factors such as the association with proteins. Contrary to the diffusion limited aggregation in dilute solutions (Li, Cui, Wang, & Yada, 2010), oat β-glucan with a lower cellotriose content has a lower tendency for spontaneous hydrogen bond association in concentrated solutions (Böhm & Kulicke, 1999b).

This type of association leads to gelling, which causes filtration and haze problems in brewing for instance. Another factor affecting the gelling intensity is molar mass. Unlike macromolecules in general, low molar mass β-glucan has a higher tendency for gelation than molecules with high molar mass ((Lazaridou, Biliaderis, & Izydorczyk, 2003); (Tosh, Wood, Wang, & Weisz, 2004)).

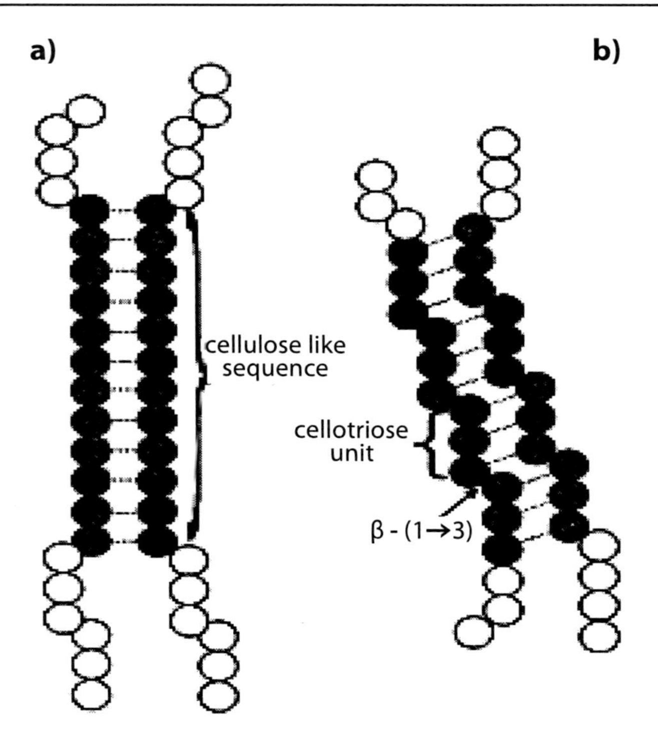

Figure 2. Chain interactions in (1→3)(1→4)-β-glucan leading to aggregates. a) The common model, according to which sequences of consecutive (1→4)-linkages interact via hydrogen bonds; b) A recently suggested model, where the hydrogen bonds form via cellotriose units (adapted with the permission of Elsevier from (Böhm & Kulicke, 1999b)).

Oat β-glucan with a molar mass of approximately $50x10^3$g/mol gelled (G'=G'') in under 2 hours in a 10% solution, while the gelation time was over 40 hours for a β-glucan with a molar mass of approximately $150x10^3$g/mol. Gels with lower a molar mass were weaker than those with higher molar mass when determined by dynamic oscillation, but stronger and less brittle when determined by static compression (Lazaridou, Biliaderis, & Izydorczyk, 2003). Beta-glucan gels typically melt at 55-65 °C, and freeze-thaw cycles promote network formation ("cryogelation"). Even relative dilute (1-4%) solutions were reported to form crosslinked, strong gels after freezing and thawing (Lazaridou & Biliaderis, 2007). In foods, gelation is affected by the matrix, interactions with other macromolecules and solutes. For example the addition of various monosaccharides significantly increased the gelation time and weakened the cryogels, but the addition of polyols promoted the gelation of high molar mass β-glucan and inhibited the network formation of low molar mass β-glucan (Lazaridou & Biliaderis, 2007)

2.3. Viscosity Properties

Due to its water solubility, high molar mass and conformation, cereal β-glucan easily forms highly viscous solutions. The logarithm of specific viscosity as a function of the logarithmic coil overlap parameter (c[η], also termed the reduced concentration) gives two or three linear regions for oat β-glucan ((Doublier & Wood, 1995); (Ren, Ellis, Ross-Murphy, Wang, & Wood, 2003); (Lazaridou, Biliaderis, & Izydorczyk, 2003); (Skendi, Biliaderis, & Lazaridou, 2003)). The regions are for dilute, semi-dilute and concentrated solutions representing the concentration-viscosity relation of certain size of β-glucan molecule. The critical concentration (c*) between dilute and semi-dilute solution represents the state when the molecules start to feel each other in the solution and the second critical concentration (c**) between semi-dilute and concentrated solutions represents the start of entanglements, after which viscosity increases more intensively with increasing concentration. Typically, as is the case of oat β-glucan, the viscosity behaviour converts from Newtonian to shear thinning when passing the critical concentration. (Doublier & Wood, 1995) first reported critical concentrations for native oat β-glucan (Mw≈1000-1500g/mol) to be c*≈ 0.1% and c**≈0.3% (c*[η]=0.7 and c**[η] =2.5). Ren et al. (2003) analysed the same samples and found one c* with 0.2-0.4% (c*[η]=2-4), and these values were followed by others as seen from the critical coil overlap parameter values c*[η] and c**[η] in table 1. The logarithmic specific viscosity of oat β-glucan has been reported to increase with a slope 1 in the dilute region, slope of 1.6-1.8 in the semi-dilute region and with a slope of 3.9 in the concentrated region as a function of the logarithmic coil overlap parameter ((Doublier & Wood, 1995); (Lazaridou, Biliaderis, & Izydorczyk, 2003)).

The flow curves and apparent viscosity values are probably dependent of the extraction method and purification degree. For example, Doublier and Wood (1995) found 0.4% unhydrolyzed oat β-glucan (Mw=1200x10^3g/mol) solution to be practically Newtonian with zero shear viscosity of approximately 70 mPa·s, while Autio et al. (1987) reported a 0.4% solution of oat β-glucan (M_w=2000x10^3g/mol) to be shear thinning with a K-value of 600 mPa·s^n, and Kivelä et al. (2009b) found a 0.2% oat bet-glucan solution (Mw≈1400x10^3g/mol, purity ≈50%, appr. 20% of proteins, no ethanol precipitation step) to be shear thinning with zero shear viscosity of 300-400 mPa·s (K≈200 mPa·s^n). Protein interactions have been proposed to increase the viscosity of oat β-glucan solutions as was shown by the lower viscosity of a trypsin-treated compared to the untreated β-glucan solution (Mälkki, 1992). Alongside of its excellent viscosity forming capacity, oat β-glucan has an excellent water-binding capacity. Oat bran was detrermined to have an approximately two-fold greater water holding capacity than wheat and barley bran (Wood P. , 1993). These properties obviously positively contributes the health functions of β-glucan, but present challenges in processing such as baking.

3. Health Functions of Oat B-Glucan

Dietary fibre is traditionally divided in two major groups: soluble and insoluble fibre. Also in case of oat β-glucan, approximately one third of the insoluble fibre of oat has been reported to consist of β-glucan, while major part of the soluble fibre is β-glucan (Manthey, Hareland, & Huseby, 1999). The differentiation is, however, more or less analytical. Physiologically the insoluble fibre is basically a bulking component, but soluble fibres may - in addition to the bulking effect - interfere with absorption of glucose, lipids and cholesterol, for instance, and have a prebiotic function in colon (Guillon & Champ, 2000).

In general, the main mechanism for the health benefits of soluble fibres is their capacity to form viscous solutions. Increased luminal viscosity in the gut modifies absorption rates, possibly by slowing mixing of the gastrointestinal contents and reducing diffusion as measured by an increase in the so-called unstirred layer adjacent to the mucosa. This cosequently delays the absorption of nutrients such as glucose, lipids and bile acids ((Story & Kritchevsky, 1976), (Vahouny, Tombes, Cassady, Kritchevsky, & Gallo, 1980)). In clinical studies, an inverse linear relationship has been demonstrated between $\log(\eta)$ or $\log(M_w{*}c)$ and postprandial glucose/insulin response and serum cholesterol levels ((Wood, Beer, & Butler, 2000); (Mäkeläinen, et al., 2007); (Regand, Tosh, Wolever, & Wood, 2009); (Wolever, et al., 2010)). In the factor $M_w{*}c$, M_w is the molar mass and c is the concentration of polysaccharide in solution. The factor is related to both intrinsic viscosity ([η]) and shear viscosity (η) by the Mark-Houwink relationship and the term $\eta \propto [\eta]c^x$. The relationship between solution viscosity and the glucose attenuating ability, serum lipid level lowering ability and immune function of cereal β-glucan has recently been comprehensively reviewed by (Wood P. J., 2010)

Despite the generally accepted mechanism, in some studies the role of viscosity have been somewhat contradictory. In addition to molar mass and extractability, other molecular characteristics, co-extracted compounds and the food matrix have been suggested to contribute to the health benefits of β-glucan ((Önning, Wallmark, Persson, Åkesson, Elmståhl, & Öste, 1999); (Kerckhoffs, Hornstra, & Mensink, 2003)).

The efficiency of native β-glucan from oat bran, grains and rolled oats in the lowering (maintaining) serum cholesterol (LDL) levels of humans is well evidenced (Ruxton & Derbyshire, 2008), but as soon as oat bran is processed to produce more sophisticated foods, behaviours of β-glucan change and the health benefits become less clear. For example, Kerckhoffs et al. (2003) suggested that the food matrix has a more relevant role than the molar mass of β-glucan, since they found no effects on serum cholesterol levels by following the consumption of 5 g of β-glucan per day in the form of yeast-leavened wheat bread enriched with oat-bran, and while oat bran consumed as a drink was effective.

Frank et al.(2004) observed that the consumption of wheat bread containing oat bran lowered the hypercholesterolemic factors of female subjects, irrespective of the molar mass of β-glucan (217 x 10^3 g/mol and 797 x 10^3 g/mol) of the consumed bread. However, neither the viscosity after extraction in physiological conditions nor the concentration of β-glucan in the extracted solutions was measured in these studies, which may affect the conclusions. Many studies have demonstrated that the extractability of β-glucan is enhanced as the molar mass decreases.

For example, Tosh et al. (2010) recently reported similar digestion viscosities *in vitro* for oat bran and extruded bran with highly degraded β-glucan in extruded products, since the extractability increased from 40% to 100% as a result of the molar mass decrease from 2200 to 210x10^3g/mol. However, the dough and bread matrix is a highly complex, and the extractability of β-glucan, even when degrading in the process, is not straightforwardly enhanced ((Johansson, Tuomainen, Anttila, Rita, & Virkki, 2007), (Rimsten, Stenberg, Andersson, Andersson, & Åman, 2003), (Moriartey, Temelli, & Vasanthan, 2010)). Some of the published clinical studies of oat products (oat gum instant whip and bread) have been failed to detect any significant reduction in serum cholesterol levels ((Beer, Wood, Weisz, & Fillion, 1997), (Törrönen, et al., 1992)). This has been suggested to be due to the poor solubility of the β-glucan in the test products used.

Biörklund et al. (2005) examined the effect of oat and barley beverages containing 5g and 10 g of β-glucan, and found a significant reduction in total serum cholesterol levels, as well as insulin and glucose levels only with the oat β-glucan beverage containing 5 g of β-glucan. The molar masses (M_w) of β-glucans were low, 70 000 g/mol in oat and 40 000 g/mol in barley, although liable to be well solubilised in a liquid product. The role of the gut viscosity in the cholesterol lowering effect was refuted by Immerstrand et al (2010), who suggested that the molecular weights and viscous properties of β-glucan in oat products may not be crucial parameters for their cholesterol-lowering effects based on their studies on cholesterol sensitive mice feed with oat bran containing a varying molar mass of β-glucan (<10 000 -2350 000 g/mol). Naumann et al. (2006) suggested that molar mass alone does not predict the cholesterol lowering effects of β-glucan, since their drinkable product with β-glucan molar mass only $80x10^3$g/mol effectively decreased the serum cholesterol concentrations in their human studies. Bae et al. (2010) noted that the bile acid binding capacity of β-glucan in rats increased after cellulase hydrolysis of β-glucan resulting in a decrease in molar mass from $1.4x10^6$g/mol to $700x10^3$g/mol.

4. Degradation of Oat B-Glucan in Food Manufacturing Processes

Processing of oat and barley to products may affect the physical and chemical properties of β-glucan, and alter its functions in the gastrointestinal track. This section focuses on the molar mass changes observed during processing, although extractability/solubility is also a fundamental factor in the functionality of β-glucan.

4.1. Baking

Baking process is one of the most studied aqueous processing of β-glucan enriched oat products. The baking process, namely fermentation and the mixing steps of the oat bran in the presence of water and wheat/rye flours degrades β-glucan ((Rimsten, Stenberg, Andersson, Andersson, & Åman, 2003); (Kerckhoffs, Hornstra, & Mensink, 2003); (Åman, Rimsten, & Andersson, 2004); (Degutyte-Fomins, Sontag-Strohm, & Salovaara, 2002);

(Andersson, Rüegg, & Åman, 2008); (Regand, Tosh, Wolever, & Wood, 2009)). The molar mass of barley β-glucan significantly and linearly decreased as the fermentation (0–90 min) and mixing (3–10 min) time of the wheat bread doughs increased (Andersson, Armö, Grangeon, Fredriksson, Andersson, & Åman, 2004). Interestingly, the rapid degradation of β-glucan resulted in three molar mass populations, which were observed after every treatment, merely shifting towards lower molar mass species with increasing fermentation time. When oat and barley concentrates were fermented with lactic acid bacteria, a significant decrease in viscosity was obtained after the fermentation, but no fermentation-induced degradation of β-glucan was detected (Lambo, Öste, & Nyman, 2005). When barley β-glucan with two different molar masses ($640x10^3$ and $210x10^3$g/mol) was added to wheat bread dough, the amount of β-glucan with a higher molar mass significantly decreased, but not that of the low molar mass β-glucan (Cleary, Andersson, & Brennan, 2007). The baking-related degradation was also slowed down by a larger particle size of oat bran (Åman, Rimsten, & Andersson, 2004), which may be explained by the lower extractability of the β-glucan from oat bran with larger particle size (Regand, Tosh, Wolever, & Wood, 2009). Muffins have been used as model products for oat bran baking with the benefit of a yeast-free process. Beer et al.(1997) reported that the muffin baking process significantly reduced the molar mass of β-glucan from $1800\text{-}1900x10^3$ g/mol to $800\text{-}1200x10^3$ g/mol depending on the recipe (with and without wheat flour and gluten) and oat bran used. However, no degradation of β-glucan in wheat flour muffins was detected by others ((Åman, Rimsten, & Andersson, 2004); (Tosh S. M., Brummer, Wolever, & Wood, 2008)). In addition to the baking processes, other aqueous processes with enzyme-active flours, such as pasta preparation, have also been reported to cause a significant degradation of β-glucan ((Åman, Rimsten, & Andersson, 2004); (Regand, Tosh, Wolever, & Wood, 2009)).Within degradation, the solubility of β-glucan has been reported to be enhanced in incubation with rye flour and rye sourdough fermentation ((Degutyte-Fomins, Sontag-Strohm, & Salovaara, 2002), in the fermentation step (Johansson, Tuomainen, Anttila, Rita, & Virkki, 2007) in muffin baking (Beer, Wood, Weisz, & Fillion, 1997), in oat crisp baking (Regand, Tosh, Wolever, & Wood, 2009) and in extrusion (Tosh S. , et al., 2010). However, the extractability/solubility of β-glucan was not found to be enhanced in association withi the significant degradation during pasta manufacturing (Regand, Tosh, Wolever, & Wood, 2009), and it decreased during baking of the yeast leavened bread (Johansson, Tuomainen, Anttila, Rita, & Virkki, 2007). The extractability of β-glucan significantly increased in muffins baked without gluten and wheat flour, but when baked with wheat flour, the extractability remained in the level of the oat bran-ingredient (Beer, Wood, Weisz, & Fillion, 1997). Processing of oat porridge by cooking is a gentle process, where the molar mass has not been found to be altered, but extractability is enhanced ((Åman, Rimsten, & Andersson, 2004); (Johansson, Tuomainen, Anttila, Rita, & Virkki, 2007); (Regand, Tosh, Wolever, & Wood, 2009)).

4.2. Freezing

The role of frozen storage in the molecular change of β-glucan has been dicussed, due to its cryogelation ability. Frozen storage has been reported to decrease the extractability, but not to change the molar mass of β-glucan ((Cleary, Andersson, & Brennan, 2007); (Kerckhoffs, Hornstra, & Mensink, 2003)). Lazaridou et al. (2003) noted that cryogelation of solutions of purified β-glucan slightly decreased the weight average molar mass of high molar mass β-glucan (1700-2000 x10^3 g/mol), but had no effect on intermediate beta-glucans (900-1400 x10^3 g/mol) and increased the molar mass of low molar mass β-glucans (60-160 x10^3 g/mol). This illustrates the complexity of the process, including aggregation but not necessarily molecular degradation. When this phenomenon is combined with the effects of the food matrix, valid information on degradation is challenging to obtain.

4.3. Processes with Mechanical Energy Input

Extrusion is an example of food process, which combines mechanical and thermal energy in conditions of controlled water activity. No kinetic study of the effect of different parameters on the physicochemical parameters of β-glucan have been reported, but the technology have been used for fragmentation purposes (Tosh S. , et al., 2010). Extrusion of oat bran with an original peak molar mass of 2500x10^3 g/mol decreased the molar mass down to 210x10^3 g/mol at temperature of 237°C, a water concentration 7%, and a standard mechanical energy of 148 Wh/kg, respectively. When the temperature was 181°C, the energy density 135Wh/kg and the water content 19%, only a slight decrease in the peak molar mass was obtained (from 2500x10^3g/mol to 2200x10^3g/mol, Tosh et al. 2010). When macaroni and flakes were produced from oat bran by extrusion at 115°C, no degradation was obtained (Åman, Rimsten, & Andersson, 2004).

A significant decrease in viscosity and molar mass was obtained during pilot plant extraction of oat gum, the manufacture of which included polytron homogenisation and pumping (Wood, Weisz, Fedec, & Burrows, 1989). Åman et al. (2004) reported an extensive decrease from 2150 x10^3 g/mol to 540 x10^3 g/mol in the molar mass of β-glucan incorporated in an apple juice, which included a pasteurisation process at 94°C for 5 s, and obviously also a homogenisation step.

When oat β-glucan with a molar mass of 200 x 10^3 g/mol was processed into an oat beverages, the molar mass decreased to 80-100 x 10^3 g/mol, whereas the manufacturing process did not affect the molar mass of β-glucan with a low original molar mass ((Naumann, van Rees, Önning, Öste, Wydra, & Mensink, 2006); (Lazaridou, Biliaderis, & Izydorczyk, 2007)). Sonication and high pressure homogenising significantly decreased the molar mass of β-glucan as discussed in section 4.4.

4.4. Instability Mechanisms in Aqueous Processing

In the presence of air and water, processing easily alters the properties of long, high molar mass polysaccharides. Water as a barrier enables the solubility and mobility of β-glucan, and also the diffusion of the substrate to enzymes, catalysts and oxidants. Biochemical and chemical reactions may degrade the polysaccharide, and energy input such as shearing, heating and radiation evidently affects the molar mass, molar shape and conformation.

4.5. Enzymatic Hydrolysis

Food processes, where beta-glucan containing ingredients are incubated in temperatures favourable for hydrolytic enzymes contain the risk of enzymatic hydrolysis of beta-glucan. Oat grains have a unique milling process that includes hydrothermal treatment to inactivate lipases and simultaneously this treatment inactivates most of the hydrolytic enzymes in oats. In baking, other baking ingredients, flours, wheat or rye flour or sourdough starters may bring hydrolytic enzymes and cause enzymatic degradation of beta-glucan in dough ((Andersson, Armö, Grangeon, Fredriksson, Andersson, & Åman, 2004); (Moriartey, Temelli, & Vasanthan, 2010)), and enzymes are also added in purpose to the dough. The dough mixing and proofing offer favorable conditions for the enzymatic hydrolysis of beta-glucan. The sourdough fermentation by pure lactic acid bacteria strains did not however cause any degradation of beta-glucan (Lambo, Öste, & Nyman, 2005), but fermentation with starters did. Startes are mixtures of various microbes that can produce beta-glucan hydrolysing enzymes. However, it has been shown that fermentation with yeast caused no additional degradation of beta-glucan in bread (Andersson, Armö, Grangeon, Fredriksson, Andersson, & Åman, 2004). Yeast contains its own beta-glucan with (1→3)- and (1→6)-β-linkages, so obviously also endogeneous glucanases, which cannot hydrolyse cereal beta-glucan.

Because of the long (1→4)- β - D -glucosyl residues that resemble cellulose, the beta-glucan chain can be hydrolysed by (1→4)- β - D -glucan endohydrolase known also as cellulase. Cellulases are probably the most common beta-glucanases in cereal flours either from microbial origin or from cereals endogenously. During grain germination cereals produce also specific (1→3,1→4)- β - D -glucan endohydrolases or lichenases that recognize the (1→3)- β -D- linkage and then hydrolyse the adjacent (1→4)-β -D- linkage (Fincher, 2009).

The cellulases and lichenases both decrease the beta-glucan molar mass and viscosity of beta-glucan in solutions, but only (1→3,1→4) -β - D -glucan endohydrolases produce short oligosaccharides in cereal beta-glucan. The cellulases are not able to cut between short (1→4)- β - D -glucosyl residues. When a viscous beta-glucan containing oat bran suspension was incubated with enzyme active rye flour the viscosity of the suspension decreased rapidly but the beta-glucan content did not change/remained stable (Degutyte-Fomins, Sontag-Strohm, & Salovaara, 2002). This showed that the degraded chains were still beta-glucan polymers and therefore beta-glucan was hydrolysed most likely by (1→4)- β - D -glucan endohydrolases i.e. cellulases and not by (1→3, 1→4) -β - D -glucan endohydrolases.

Many cereal processes make use of food grade enzymes from malt (germinated grains), bacterial or fungal origin. When using these enzymes it is likely that several activities of enzymes including cellulases or lichenases can be obtained at the same time.

From the process point of view enzymatic hydrolysis is part of the beneficial quality effects in the process of cereal food products like the hydrolysis of cell walls of barley in brewing or softening of the bran particles in high fibre baking. Enzymes acting on beta-glucan are also utilized in oat bread baking, where the high water binding capacity and viscosity causes challenges in bread texture.

4.6. Thermal Degradation

Heating of aqueous materials is commonly used for the pasteurization, sterilization and modification of foods at temperatures from 90-120 °C. Heating influences polysaccharides by accelerating their molecular vibration, collisions and chemical reactions in solutions resulting in side-group elimination, random scission (fragmentation) or depolymerisation of the polysaccharide chain (Pielichowski & Njuguna, 2005). The degradation rate and products depend on the mechanisms, which are altered by the temperature range, processing time, pressure, and solution properties. The heat-induced hydrolysis of glycosidic bonds is catalysed by OH^- -ions and H_3O^+-ions (alkali and acid hydrolysis), and thermal degradation is thus strongly pH-dependent. In addition to hydrolysis, oxidation (Robert, Barbati, Ricq, & Ambrosio, 2002) and Maillard reactions (Davídek, Robert, Devaud, Arce Vera, & Blank, 2006) may contribute to the chain cleavage of polysaccharides at elevated temperatures, which makes the solvent quality an important parameter. The concentration and molecular properties of polysaccharides may also affect the rate of thermal degradation.

The sensitivity of polysaccharides to thermal degradation has generally been estimated by the Arrhenius equation (Eq. 1).

$$k = A\exp\left(\frac{-Ea}{RT}\right)$$

where k (s^{-1}) is first order rate constant for the degradation process at temperature T (K) and with a constant A (s^{-1}), Ea (J/mol) is the activation energy, meaning the lowest energy load needed to initiate the reaction, and R is the gas constant. The Arrhenius equation may be used for the results of the thermogravimetric (TG) method or the degradation constant (k) can be estimated from the change in molar mass during thermal treatments as done for instance by Bradley and Mitchell (1988):

$$\frac{1}{Mw,t} - \frac{1}{Mw,0} = \frac{k}{2m} t$$

where M_{wt} and M_{w0} are the molar masses of β-glucan after heat-treatment for time *t* and 0, respectively, and m is the molar mass of the monosaccharide.

Lai et al. (2000) pooled activation energy data from various studies on common water-soluble polysaccharides, which were heated at 25°C-160^{0}C (1% solutions) in atmospheric conditions, and observed a linear correlation between A and E_a. This implies an isokinetic relationship between the different food-related carbohydrates and indicates that they all follow similar reaction mechanism despite the differences in side chains, charges or molecular size. Typically the activation energy values were found to be near to 100 kJ/mol, for example 106 kJ/mol and 97 kJ/mol for agarose and κ-carragenan, respectively, at 95°C and 80 kJ/mol for carboxymethylcellulose (40-110^{0}C) ((Lai, Lii, Hung, & Lu, 2000); (Bradley & Mitchell, 1988)). This indicates random scission acting on the polysaccharide backbone (Soldi, 2005). However, the thermal degradation of polymeric material generally follows more than one mechanism and several models have been applied to estimate the kinetics (Pielichowski & Njuguna, 2005). In the case of water-soluble food thickeners, generalising is also difficult. For example, a relatively good thermal stability, defined as the viscosity change after 60 min treatment at 120°C, was reported for carboxymethylcellulose compared to alginate, xanthan and guar gum (Mitchell, Reed, Hill, & Rogers, 1991). In addition, two galactomannans, locust bean gum and guar gum, were reported to differ from their activation energies having E_a=98 kJ/mol and 63 kJ/mol, respectively, at 70^{0}C-121^{0}C (Kök et al., 1999). In the case of oat β-glucan, the backbone was shown to be cleaved at 120°C ((Wang, Wood, & Ross-Murphy, 2001), (Kivelä, Sontag-Strohm, Loponen, Tuomainen, & Nyström, 2011a)), whereas only the degree of aggregation of xyloglucan and dextran was affected under the same conditions (Wang, Wood, & Ross-Murphy, 2001). In other study, viscosity of oat β-glucan solution was relatively stable against the heating at 100°C (Autio, Myllymäki, & and Mälkki, 1987).

When solutions of oat β-glucan were heated at 120^{0}C for 15 minutes in a β-glucan extract (purity of ≈50%), and after purification steps (purity≈85%), the molar mass decreased from 1600 x 10^3g/mol to 720 x 10^3g/mol in the unpurified and to 1400 x 10^3g/mol in the purified solution of β-glucan (Kivelä, Henniges, Sontag-Strohm, & Potthast, 2011b). This highlights the complexity of food matrices and the susceptibility of β-glucan to processes in its natural environment. In the presence of proteins and metals, for instance, chemical reactions such as oxidation and Maillard reactions may take place and promote the fragmentation of beta-glucans.

The same process is illustrated in Figure 3, where solutions of highly purified commercial oat β-glucan (OBG, purity≈99%, M_w≈250x10^{3}g/mol, c=1%) and native oat β-glucan (NBG, purity≈50%, M_w≈1600x10^{3}g/mol, c=0.5%) in the presence of co-extracted compounds (proteins, minerals, phytates) were heated at 95°C and 120°C.

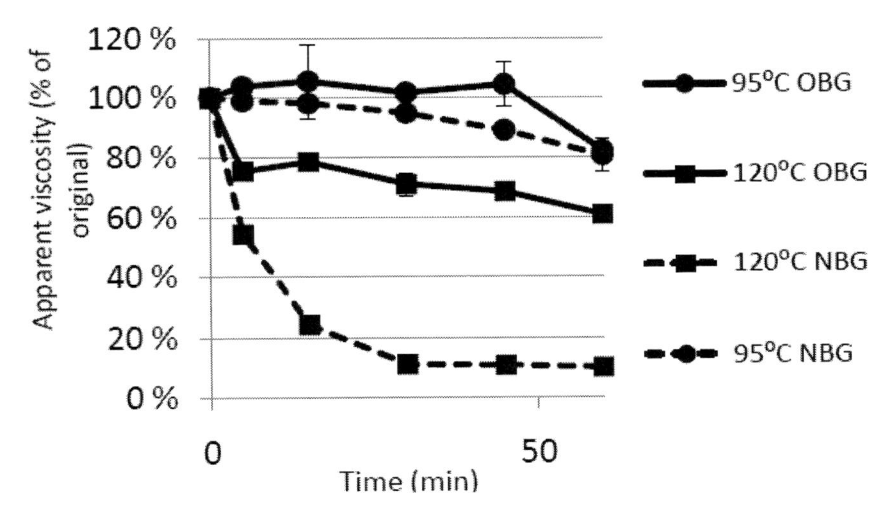

Figure 3. The decrease in relative apparent viscosity (% of original) of native (NBG) and highly purified (OBG) oat beta-glucan as a function of time (h) when heating at 95 °C and 120 °C (unpublished data of Kivelä/University of Helsinki.

The thermal degradation of native β-glucan was highly intensive at 120°C, while the purified remained relatively stable ((Kivelä, Sontag-Strohm, Loponen, Tuomainen, & Nyström, 2011a); (Kivelä, Henniges, Sontag-Strohm, & Potthast, 2011b)). The decrease in the intrinsic viscosity of agarose and κ-carragenan when heated at 75°C, 85°C and 95°C is illustrated in Figure 4A and 4B, respectively. Autio et al. (1987) reported that the viscosity of 0.7% oat β-glucan solutions decreased to a tenth of the original level in 145 hours when heated at 100°C. A similar viscosity decrease calculated from Figure 3 is gained in approximately 5 hours at 95°C for the two different oat β-glucans, in approximately 3 hours at 120°C in the case of purified β-glucan and in 20 minutes at 120°C in the case of β-glucan extract.

4.6.1. Acid Hydrolysis

Acid hydrolysis is a widely utilised degradation reaction in the fragmentation and modification of polysaccharides, and is also used in the fragmentation of β-glucan in addition to enzymatic hydrolysis. Essentially, oxonium ions (H_3O^+) catalyse the hydrolysis of glycosidic linkages of polysaccharides, usually induced by elevated temperatures.

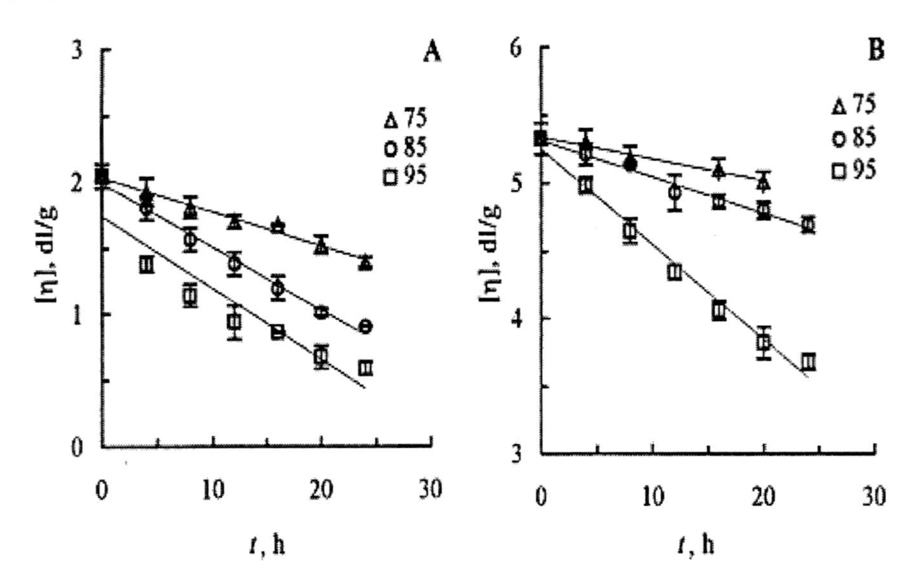

Figure 4. The decrease in the intrinsic viscosity [η] of A) agarose and B) κ-carragenan as a function of time (h) when heating at 75–95 °C (adapted with the permission of Elsevier from (Lai, Lii, Hung, & Lu, 2000)).

The proton of the catalysing acid rapidly interacts with the glycosidic oxygen in the linkage, which is followed by a rapid addition of water-molecules to the intermediate, carbonium cation, thus resulting in the cleavage of the glycosidic bond and stable hydrolysis products.

The rate and the products of acid hydrolysis are affected by the type of acid, its concentration, pH, temperature, pressure and also the molecular characteristics of the polysaccharides such as their solubility, structure and conformation. In mild conditions, polysaccharides chains will be randomly cleaved resulting in a decrease in the weight average molar mass. In stronger conditions, hydrolysis results in oligo- and monosaccharides and their derivatives. Typical acid hydrolysis products of glucans among numerous minor products include carboxylic acids such as formic, acetic and lactonic acid, and hydroxymethylfurfural (HMF). Acid hydrolysis of oat β-glucan performed with 0.1 M HCl at 70°C for 30-90 min resulted in fragments with M_w-values from 1200x10^3 g/mol to 30-170 x10^{3}g/mol (Tosh, Wood, Wang, & Weisz, 2004). The same acid conditions were investigated at temperatures 37°C (5h and 12 h), 70°C (5h and 12h) and 120°C (1h) (Johansson, Virkki, Anttila, Esselström, Tuomainen, & Sontag-Strohm, 2006). The stomach mimicking conditions (37°C) only slightly affected the flow properties, but no formation of oligo- or monosaccharides occured. At 70°C glucose and cellobiose were formed in significant amounts and at 120°C, the recovery of glucose was ≈100% and that of cellobiose ≈5%.

Most typically, the acid hydrolysis of polysaccharides follows first order kinetics as shown with the linear, neutral and water-soluble polysaccharide galactoglucomannan as a strong and linear relation between the reciprocal molar mass and reaction time (Wang, Ellis, & Ross-Murphy, 2000). However, the branched water-soluble polysaccharides xanthan and scleroglucan exhibited a two-stage acid hydrolysis beginning with a slow-rate stage, and followed by a second stage in which the apparent degradation rate was much higher (Hjerde, Kristiansen, Stokke, Smidsrod, & Christensen, 1994). The acid hydrolysis of barley β-glucan ($M_{w,0}$=140 000 g/mol) by H_3PO_4 required a certain time at pH 2.5-4.5, before the degradation began based on the results of Vaikousi and Biliaderis (2005) (Figure 5), and thus the hydrolysis might not have followed first order kinetics. The ordered structures of κ-carragenan were shown to hydrolyse with a slower rate than disordered structures (Hjerde, Smidsrød, Stokke, E., & Christensen, 1998), which may also reflect the kinetics of the small molar mass barley β-glucan with a high tendency for side by side aggregation as discussed in section 1.2.

Acid hydrolysis of β-glucan has been suggested to play a role in beverages, the manufacture of which usually includes heat treatments in acidic conditions derived from fruit and berry juices. When an oat bran concentrate was processed into a drinkable product with apple juice, the molar mass of β-glucan decreased from 2100x10^{3}g/mol to 600x10^{3}g/mol (Åman, Rimsten, & Andersson, 2004). In addition, molar mass of oat β-glucan in a blackberry juice or apple/pear flavoured oat drink has been reported to be 80x10^{3}g/mol ((Lazaridou, Biliaderis, & Izydorczyk, 2007); (Naumann, van Rees, Önning, Öste, Wydra, & Mensink, 2006)). The processing of beverages usually includes a heating step, which may lead to acid hydrolysis. However, the heating step in pasteurization and sterilization is few seconds with effective cooling, and pH of regular juices is approximately 3-3.5, although it is likely to be even higher in cereal drinks.

Under these pH-conditions, the viscosity of soluble barley β-glucan ($M_{w,original}$=250 000 g/mol) was not significantly affected by 5 minutes of treatment at 82.5°C (Figure 5, (Vaikousi & Biliaderis, 2005)). In addition, when the effects of the most abundant carboxylic acids (malic, citric and ascorbic acid) present in equal concentrations in fruit and berry juices were investigated in an oat β-glucan solution under mildly acidic conditions (pH 4.1-4.4) with heating for 5 minutes at 90^{0}C, no acid hydrolysis-dependent effect was detected (Kivelä, Nyström, Salovaara, & Sontag-Strohm, 2009b). A slight decrease in the samples with malic acid and HCl (pH-control) was obtained, but a similar viscosity decrease was also obtained in the extract with pH of 7. Similar acid-induced effect was obtained in the case of galactoglucomannan by Wang et al. (2000), and this slight change probably reflected changes in the shape of the aggregates, their assemblies or in the molecules due to the pH change, and not hydrolysis of the backbone.

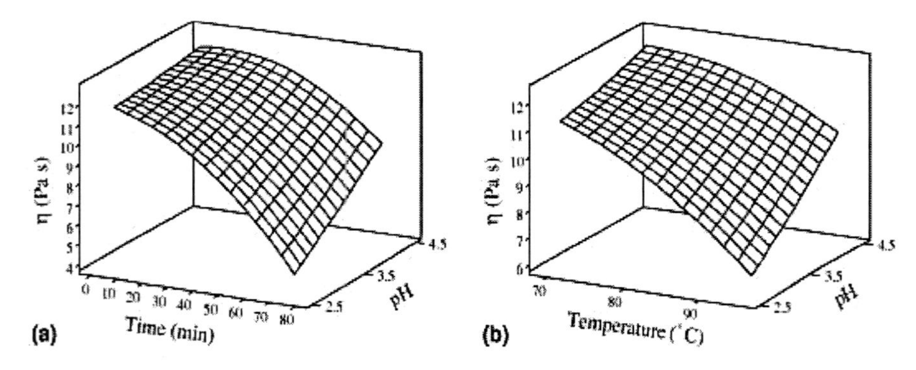

Figure 5. Decrease in the apparent viscosity of barley beta-glucan solution (4%, w/w) when treated with H_3PO_4 A) at a constant temperature (82.5 °C) or B) for a constant hydrolysis time (39 min) (adapted with the permission of Elsevier from (Vaikousi & Biliaderis, 2005)).

Citric acid significantly decreased the viscosity, and interestingly ascorbic acid drastically decreased the viscosity ((Kivelä, Gates, & Sontag-Strohm, 2009a), (Kivelä, Nyström, Salovaara, & Sontag-Strohm, 2009b)). The extent of the decrease in viscosity was also caused by ascorbic acid (2-4mM, 35-70mg/100ml) in a neutral pH and at room temperature (Kivelä et al., 2009b), and the effect was concluded to be due to oxidation initiated by ascorbic acid as discussed in section 4.2.2. In the case of drinkable β-glucan products, the mechanical energy applied in homogenisation and pumping also easily alters the soluble β-glucan.

4.6.2. Degradation under Alkaline Conditions

In foods or food manufacturing, alkaline conditions are rare. In analytics, alkaline conditions, usually created with sodium hydroxide, have been used to enhance the extraction recovery of β-glucan from the flours ((Dawkins & Nnanna, 1995); (Beer, Wood, Weisz, & Fillion, 1997); (Burkus & Temelli, 1998)). Barium hydroxide has been used to enhance the purity and extraction yield of the extracted β-glucan, which may be associated with water-insoluble arabinoxylan (Brummer;Jones;Tosh;& Wood, 2008). In addition, sodium hydroxide has been used to examine the aggregation behaviour of β-glucan due to its ability to reveal aggregates as demonstrated by light scattering studies (Li, Wang, Cui, Huang, & Kakuda, 2006).

One of the basic principles of the complex carbohydrate degradation in alkaline conditions is the keto-enol tautomerism of monosaccharides. At elevated temperatures, hydroxyl ion attacks the enediol anion, which then undergoes β- elimination i.e. the removal of the hydroxyl/alkoxyl group in the β-position to the negatively charged oxygen ((Whistler & BeMiller, 1958); (Knill & Kennedy, 2003)). In an alkaline environment, several degradation reactions takes place, the most important being peeling and stopping reactions, and alkaline hydrolysis. The peeling reactions start from the existing protonated end group (reducing end) cleaving the glycosidic bonds by β-alkoxycarbonyl elimination by one monomer at a time. Peeling is the dominant reaction resulting in molar mass losses at temperatures <170°C in an alkaline environment, but at temperatures above 170°C alkaline hydrolysis reactions starts to dominate (Nevell, 1985). In hydrolysis, new reducing end-groups are formed, which are also new subjects of the endwise degradation i.e. peeling. Knill and Kennedy (2003) listed products of alkaline degradation of glucose, cellobiose and cellulose within a temperature range of 20°C-200°C by several bases. Organic acids such as formic, acetic, glycolic, lactic and saccharinic acids were formed in significant amounts under all the conditions from all the materials. Compared to the cellulosic material, similar compounds of organic acids also formed in the alkaline hydrolysis of starch, although at a slower rate (Krochta, Tillin, & Hudson, 1987).

Some indications exist of the degradation or depolymerisation of β-glucan in alkaline conditions. Oat β-glucan declined in peak molar mass from 1800-1900x10^3g/mol to 1200x10^3g/mol when it was extracted from an oat bran with 1.3M NaOH (16 h) instead of hot water (2h, 90°C) or when extracted under physiological conditions (Beer, Wood, Weisz, & Fillion, 1997). Alkaline extraction conditions (pH 8-10, 55°C) also resulted in reduced extraction yields of barley β-glucan, which were associated with alkaline depolymerisation ((Temelli, 1997); (Symons & Brennan, 2004)). However, there are several studies about the NaOH-extraction, where no degradation or yield losses have been observed. Brummer et al. (2008) reported that saturated barium hydroxide [$Ba(OH)_2$] drastically cleaved β-glucan. The degree of degradation at room temperature was dependent on the time and concentration of $Ba(OH)_2$. However, in this study the effect of alkalinity in barium hydroxide-derived degradation was expected to be excluded by using sodium hydroxide as a control (Brummer;Jones;Tosh;& Wood, 2008). The peak molar mass of oat β glucan after aqueous extraction was $1200x10^3$g/mol, while after 17 hours of extraction with NaOH it was $1000x10^3$g/mol and after 17 hours of extraction with $Ba(OH)_2$ $70x10^3$g/mol. Based on this considerable difference, the authors suggested the degradation to be associated with barium derived oxidation rather than alkaline scission. However, barium hydroxide was among the first compounds reported to produce saccharinic acids from the alkaline degradation of glucose as reviewed by Knill and Kennedy (2003).

One of the most critical aspects of alkaline extraction and the analysis conditions concerns with processed β-glucan-enriched products. If the β-glucan chain is oxidized in processing, the alkaline analysis conditions may alter the information obtained on process-related molar mass-changes. This is because the β-elimination reaction may start not only from the reducing ends, but also from the carbonyl groups ((BeMiller, 2007), scheme 2), which are introduced to the chain under oxidation as discussed in section 4.2.2. The carbonyl groups may occur randomly in the chain, and the β-elimination reaction easily leads to a steep decrease in the weight average molar mass as shown with cellulose of cotton linters (Figure 6, (Potthast, Rosenau, & Kosma, 2006)). The effect was obtained by relatively mild alkaline conditions (pH 11 produced with NaOH, 40^0C) of a pre-oxidized cellulose molecule.

The Figure 6 illustrates the disappearance of the carbonyl groups consistently with a molar mass decrease indicating OH^--ions induced β-elimination. Similar effect was shown by oxidized xyloglucan, as the viscosity of its solutions irreversibly decreased by adjusting the pH from 4.7 to 12.7 and back to 4.7 by using NaOH ((Miller & Fry, 2001)). If covalent linkages had not been cleaved, the neutralisation would have reformed the weak associations and returned the viscosity as also seen with oat β-glucan (Wood, Weisz, Fedec, & Burrows, 1989).

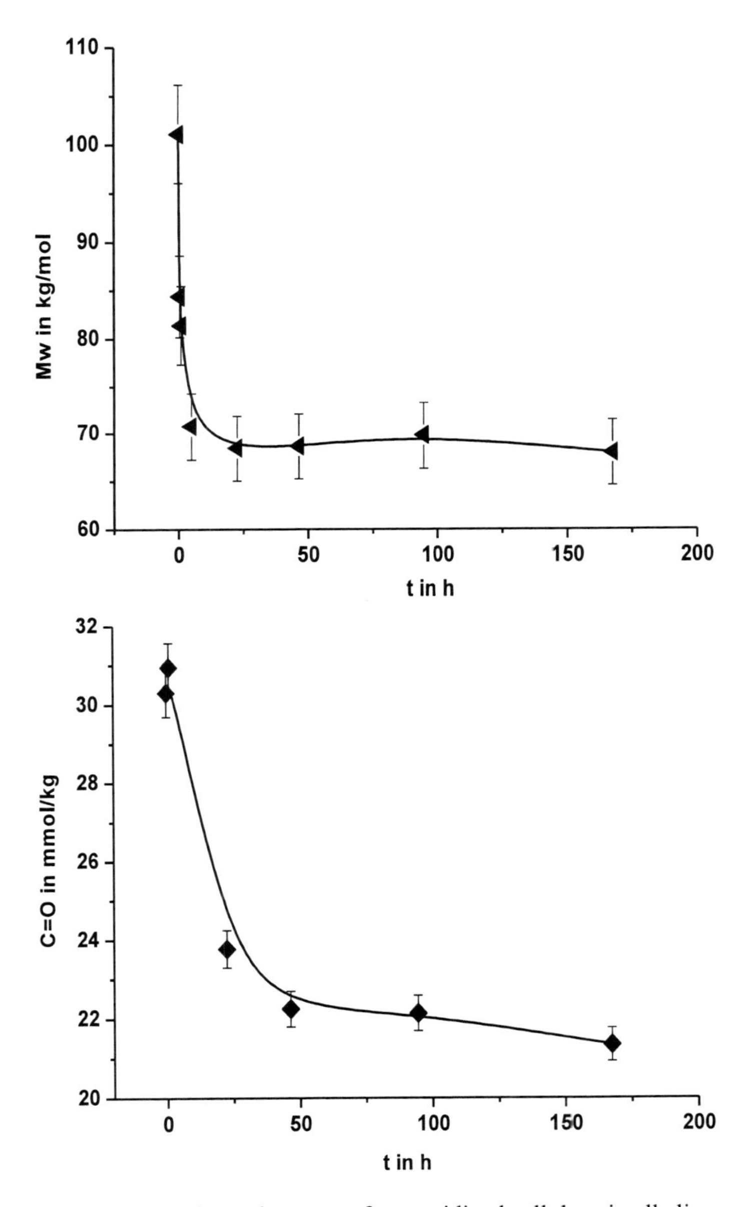

Figure 6. A) Decrease in the molar mass of pre-oxidized cellulose in alkaline conditions (pH 11, 40°C), B) The consistent decrease in carbonyl groups formed in oxidation (adapted with the permission of Springer from Potthast et al., 2006).

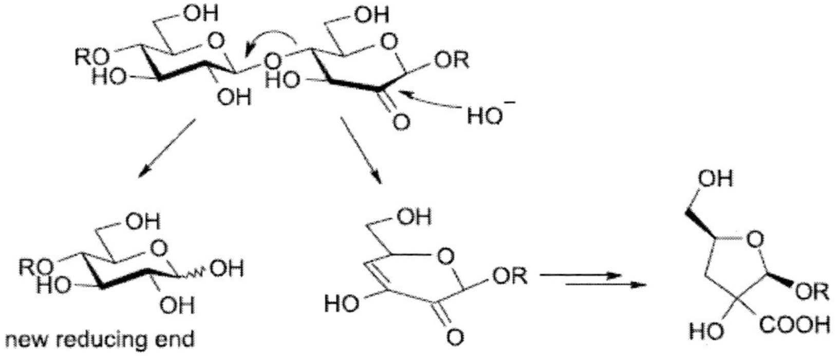

Scheme 2. Cleavage of the glycosidic bond of a glucane chain by β-elimination, when the carbonyl group exist in C2. The reaction products formed by β-elimination differ significantly depending on the position of the carbonyl group. Drawn after Potthast et al. (2006).

4.7. Oxidative Cleavage

The presence of molecular oxygen accelerates the degradation rate of polysaccharides in various conditions and is critical for the initiation of degradation. For example, teh activation energies of the thermal degradation of cellulose, corn starch and chitosan increased from 59 kJ/mol, 50 kJ/mol and 160 kJ/mol to 242 kJ/mol, 470 kJ/mol and 181 kJ/mol, respectively, as the atmospheric conditions were replaced by nitrogen ((Peniche-Covas, Argüelles-Monal, & San Román, 1993); (Aggarwal, Dollimore, & Heon, 1997)).

Molecular oxygen, hydrogen peroxide and ozone have also been utilized with metal catalysts in carbohydrate modification based on the oxidation bathways ((Arts, Mombarg, van Bekkum, & Sheldon, 1997); (BeMiller, 2007)). Also radiation, sonication and supercritical water are widely utilised in oxidative modification of several constituents.

In general, oxidation is initiated under conditions where alkyl- and hydroperoxyl radicals are formed in reactions between molecular oxygen and organic compounds, or in the decomposition of organic compounds ((Robert, Barbati, Ricq, & Ambrosio, 2002)). The propagation reactions produce strong oxidants, including reactive oxygen species (ROS). The ROS most generally referred to are the superoxide anion ($O_2^{\bullet-}$) and its protonized form the hydroperoxyl radical ($HO_2^{\bullet}$), hydrogen peroxide (H_2O_2) and hydroxyl (and alkoxyl) radicals (•OH and •OR), the hydroxyl radical being the most reactive in the order H_2O_2 < $O_2^{\bullet-}$ < •OH (Halliwell & Gutteridge, 2007). The Hydroxyl/alkoxyl radicals are produced by homolytic scission of the O-O bond of hydrogen peroxide (H_2O_2) or organic peroxides (ROOH), which may be initiated by metals, heat and radiation. One of the widely accepted source of hydroxyl radicals in biological systems is the Haber-Weiss cycle (reactions 1-3) including the Fenton reaction (3) (Wardman & Candeias, 1996)

$Cu^{+}/Fe^{2+}+O_2 \leftrightarrow Cu^{2+}/Fe^{3+}+O_2^{\bullet-}$ (Reaction 1)

$2O_2^{\bullet-}+2H^{+} \rightarrow H_2O_2+O_2$ (Reaction 2)

$Cu^{+}/Fe^{2+} + H_2O_2+H^{+} \rightarrow Cu^{2+}/Fe^{3+}+H_2O + {}^{\bullet}OH$ (Reaction 3, Fenton)

In the Fenton reaction, hydrogen peroxide is reduced in a transition metal catalyzed reaction, which produces hydroxyl radicals. However, iron oxidation chemistry is complex in biological solutions and the role of Fenton chemistry may have been over-estimated. For example, iron can complex with molecular oxygen (forming a ferryl ion or a perferryl ion), chelates or other molecules such as phenols and proteins and may form oxidative non-radical compounds ((Kaneda, Kano, Osawa, Ramarathnam, Kawakishi, & Kamada, 1988); (Welch, Davis, & Aust, 2002); (Qian & Buettner, 1999))

Hydroxyl radicals non-selectively attack macromolecules at diffusion and pH-dependent rates ((von Sonntag, 1980); (Gilbert, King, & Thomas, 1984)). The radicals attack the molecules in their immediate vicinity, since their life time is ≈1 ns. This leads the localisation of metals being one of the dominant factors in the oxidative damage of macromolecules and may increase the sensitivity of high molar mass polysaccharide to oxidation as they have been reported to associate with metals ((Platt & Clydesdale, 1984); (Chevion, 1988)). The attack initiates chain reactions by abstracting the hydrogen from C-H moieties, which leads to the formation of an alkoxyl radical. The alkoxyl radical, which is a radicalised chain in the case of polysaccharides, can react with atmospheric oxygen leading to a peroxyl radical intermediated carbonyl group formation as shown in scheme 2. In the case of polysaccharides, depending on which carbon is attacked, this can lead to the formation of new functional groups of the glucose residues (→carbonyl→carboxyl), the cleavage of the glycosidic bond, and/or the formation of numerous degradation products such as lactones (Arts, 1997; Potthast et al., 2006).

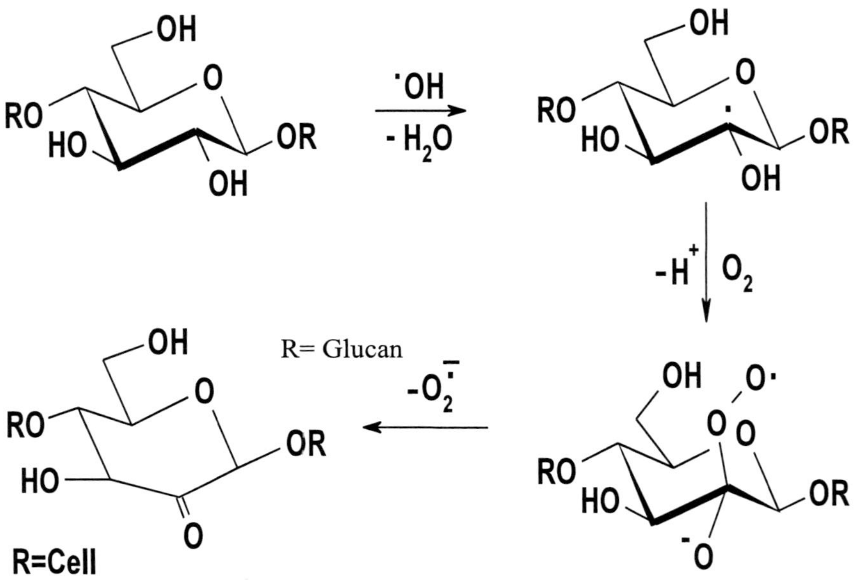

Scheme 3. Carbonyl group introduction by hydroxyl radical (·OH) attack on C2.

Hydrogen abstraction is non-specific in a glucose ring of neutral polysaccharides, but attack on the carbons involved in the glycosidic bonds may be hindered. However, the unpaired electrons may rearrange in the glucose residues or the conditions may promote β-elimination, and the glycosidic bonds are thus vulnerable to oxidative cleavage ((Gilbert, King, & Thomas, 1984), (Potthast, Rosenau, & Kosma, 2006)). The polysaccharide radicals (reaction 4) decay either intramolecularly with another radical situated in the same chain or intermolecularly with another polymeric or non-polymeric radical (von Sonntag, 1980).

Oat β-glucan has been shown to oxidize in aqueous matrices ((Kivelä, Gates, & Sontag-Strohm, 2009a), (Kivelä, Nyström, Salovaara, & Sontag-Strohm, 2009b), (Kivelä, Sontag-Strohm, Loponen, Tuomainen, & Nyström, 2011a), (Kivelä, Henniges, Sontag-Strohm, & Potthast, 2011b)). As mentioned above, ascorbic acid (1-10 mM= 18-180mg/100ml) drastically decreased the viscosity of oat β-glucan solutions at room temperature, while other organic acids had no significant effect. Ascorbic acid, or ascorbate when it occurs in the pH range 4.2-11.6, is a widely used antioxidant, based on relatively stable ascorbyl radicals, which scavenge the oxidative free radicals. However, ascorbic acid is also an excellent reducing agent (thus used as flour improver), and may behave as a pro-oxidant in aqueous solutions in the presence of transition metals ((Buettner & Jurkiewicz, 1996); (Fry, 1998)). It can reduce and stabilise transition metals such as iron into the reduced form (reaction 4) and reduce dissolved oxygen to hydrogen peroxide (reaction 5), which can further produce oxygen radicals by the Fenton reaction (reaction 3).

$AH_2 + 2\ Cu^{2+}/Fe^{3+} \rightarrow A + 2\ H^{+} + 2\ Cu^{+}/Fe^{2+}$ (Reaction 4)

$AH_2 + O_2 \rightarrow A + H_2O_2$ (Reaction 5)

$Cu^{+}/Fe^{2+} + H_2O_2 \rightarrow OH^{\bullet} + OH^{-} + 2H_2O + Cu^{2+}/Fe^{3+}$ (Reaction 3, Fenton)

In addition, ascorbic acid may stabilise hydrogen peroxide, which easily decomposes under atmospheric conditions (Arts, Mombarg, van Bekkum, & Sheldon, 1997). In addition to cereal β-glucan, the pro-oxidative effect of ascorbic acid has been shown as a viscosity decrease for xyloglucan, pectin, carboxymethyl cellulose, dextran, cassava starch and chitosan solutions ((Vallès-Pàmies, Barclay, Hill, Mitchell, Paterson, & Blanshard, 1997); (Fry, 1998); (Zoldners, Kiseleva, & Kaiminsh, 2005)), and as an accelerated free radical formation in beer (Andersen;Outtrup;& Skibsted, 2000). In addition, ascorbate contributes, together with metals to oxidative stress and DNA-damage, which are the most studied fields of hydroxyl radical attack ((Valko, Morris, & Cronin, 2005)). Other reducing agents such as glutathione have also been reported to degrade polysaccharides such as cassava starch (Vallès-Pàmies, Barclay, Hill, Mitchell, Paterson, & Blanshard, 1997). The ascorbic acid induced viscosity loss of β-glucan solutions was shown to slow down when the hydrogen peroxide decomposer catalase, or the hydrogen peroxide scavengers mannitol and glucose were present (Figure 7) , which supports the role of oxidation (Kivelä, Nyström, Salovaara, & Sontag-Strohm, 2009b). In purified β-glucan solution, the inhibition of glucose was almost total and the depletion of oxygen significantly decreased the rate of ascorbic acid induced degradation (Kivelä, Gates, & Sontag-Strohm, Degradation of cereal beta-glucan by ascorbic acid induced oxygen radicals, 2009a). The molar mass of β-glucan decreased from 1400×10^3 g/mol to 360×10^3 g/mol after 4 hours of treatment with 1 mM ascorbate at room temperature and the conformation of degraded β-glucan remained as a random coil, as the Mark-Houwink exponent was in the range 0.6-0.7 in the ascorbic acid treated samples (Kivelä,

Henniges, Sontag-Strohm, & Potthast, 2011b). This indicates that the viscosity loss was caused by cleavage of the backbone of the β-glucan molecule.

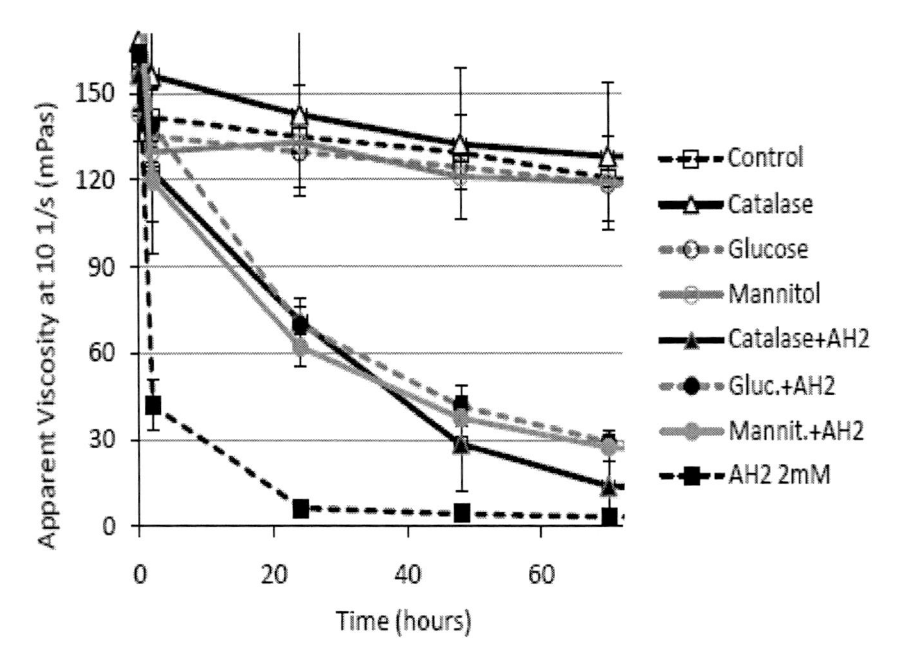

Figure 7. Decrease in the apparent viscosity of an oat beta-glucan extract (0.2%) when treated with ascorbic acid, and inhibition by hydroxyl radical scavengers (glucose and mannitol) and a hydrogen peroxide decomposer (catalase) of the ascorbic acid-induced viscosity loss. The pH was 4.5 ± 0.2 and temperature +6–10 °C during the experiments (drawn from the data of Kivelä et al., 2009b).

Oxidation was also found to contribute to the thermal degradation of oat β-glucan in solution, as illustrated in Figure 3, since the formation of strongly oxidative radicals, most likely hydroxyl and alkoxyl radicals, correlated with the heat -induced molecular degradation of β-glucan (Kivelä, Sontag-Strohm, Loponen, Tuomainen, & Nyström, 2011a).

Oxidation induced not only viscosity loss but also gel formation in relatively dilute solutions of barley β-glucan (0.7%, ascorbate), guar galactomannan (1%) and xyloglucan (1%) ((Kivelä, Gates, & Sontag-Strohm, 2009a); (Parikka;Leppänen;Pitkänen;Reunanen;Willför;& Tenkanen, 2010)). The galactomannan and xyloglucan were enzymatically oxidized with galactoxidase by targeting their primary alcohols in C6. When the carbonyls are introduced in the backbone, they can form hemiacetal crosslinks with hydroxyl groups of the glucose residues. The crosslinking results in molar mass increase as seen by thermal treatments of pullulans and by periodate treatments of cellulose ((Strlic, Kocar, Kolar, Rychlý, & Pihlar, 2003); (Potthast, Kostic, Schiehser, Kosma, & Rosenau, 2007)).

Kivelä et al. (2011b) obtained a low and a high molar mass population after heating highly purified oat β-glucan at 120°C for 30 minutes. The original population was in order of 10^5 g/mol, the low molar mass population in order of 10^4g/mol and the high molar mass population in order of 10^7g/mol, and heat-induced carbonyl group formation was detected, the phenomenon was, thus, concluded to be due to hemiacetal crosslinking by the oxidized β-glucan molecules. In addition to the backbone cleavage, gelation also affects the functionality of cereal β-glucan and may contribute to the physiological activity by lowering the extractability as shown by freeze-thawed β-glucan muffins (Tosh S. M., Brummer, Wolever, & Wood, 2008) or by improving the protective characteristics of the β-glucan by strengthening the unstirred layer in the lumen.

Oxidation of oat β-glucan was recently reported to significantly promote the decrease in the levels of triglyceride, total cholesterol, LDL-C and VLDL-C in hypercholesterolemic rats (Park, Bae, Lee, & Lee, 2009). This was explained by an enhanced water solubility and in vitro bile acid binding capacity, due to the acidic carbonyl groups introduced to the C6 of the glucose residues by targeted TEMPO-oxidation. Thus, strongly depending on the oxidants and conditions, the oxidation may be a challenge or an advantage in the development of β-glucan products.

4.7.1. Managing the Oxidation of β-Glucan in Aqueous Processing

The oxidation reactions in foods are generally managed by antioxidants, chelating agents and oxygen depletion. In the case of polysaccharide oxidation, antioxidants seem to have a controversial role in water solutions/ dispersions.

Not only ascorbic acid, but also for example sulphite easily degradred soluble polysaccharides such as starch and oat β-glucan ((Paterson, Hill, Mitchell, & Blanshard, 1997); (Kivelä, Sontag-Strohm, Loponen, Tuomainen, & Nyström, 2011a)). However, condition-dependently, sulphite also protected the polysaccharides from oxidation when used together with propyl gallate or ascorbic acid, and the thermal degradation of various polysaccharides, including oat β-glucan, was inhibited or significantly slowed down ((Hill & Gray, 1999); (Kivelä, Sontag-Strohm, Loponen, Tuomainen, & Nyström, 2011a)). Sulphite is traditionally referred to as an oxygen scavenger, whereas propyl gallate and ascorbic acid are free radical scavengers, thus the different mechanism may create the synergism obtained.

Chelates also have a controversial effect on polysaccharide degradation. For example EDTA accelerates hydroxyl radical generation via the Fenton reaction, but DTPA and phytate may delay the oxidation cycle (Graf, Empson, & Eaton, 1987). Water has a major role in degradation processes, and β-glucan is obviously more susceptible to degradation in aqueous matrices.

For example, physical hindrance, studied in soft starch gels (G'≈10Pa), significantly slowed down the rate of ascorbic acid induced β-glucan degradation (Kivelä, Nyström, & Sontag-Strohm, 2008). This may also be true in baking, for instance, where free radicals have hindrances and various substrates to attack.

4.8. Mechanical Energy Input

Effect of shear and mechanical stress on the properties of oat β-glucan has been studied during shearing, sonication and homogenisation processes of β-glucan solutions ((Izydorczyk & Biliaderis, 2000); (Kivelä, Pitkänen, Laine, Aseyev, & Sontag-Strohm, 2010)). Sonication clearly reduced the molar mass of β-glucan when barley grains or β-glucan solutions were treated ((Izydorczyk & Biliaderis, 2000); (Böhm & Kulicke, 1999a)). No permanent viscosity or molar mass loss was obtained when high shear rates (24 000 rpm, 16 000 s^{-1}) were applied to β-glucan solutions ((Wood, Weisz, Fedec, & Burrows, 1989); (Vaikousi & Biliaderis, 2005)), but high pressure homogenisation (shear rates in order to 1×10^6-1×10^7 s^{-1}) reduced molar mass and viscosity of oat β-glucan extracts (Kivelä, Pitkänen, Laine, Aseyev, & Sontag-Strohm, 2010). The microfluidizer treatment decreased the weight average molar mass from 1400×10^3 g/mol to 300×10^3g/mol after 10 seconds and to 150×10^3g/mol after 5 minutes treatment with 100MPa (Figure 8). Homogenisation, particularly microfluidizing has been demonstrated to be an effective fragmentation technique also for high molecular weight chitosan ($2000 * 10^3$g/mol), xanthan ($25 * 10^3$g/mol) and gum tragacanth ($850 * 10^3$g/mol) ((Kasaai, Charlet, Paquin, & Arul, 2003); (Lagoueyte & Paquin, 1998) and (Silvestri & Gabrielson, 1991)), and high pressure valve homogenisation has been revealed to degrade polysaccharides such as methyl cellulose and modified starch ((Floury, Desrumaux, Axelos, & Legrand, 2002); (Nilsson, Leeman, Wahlund, & Bergenståhl, 2006)).

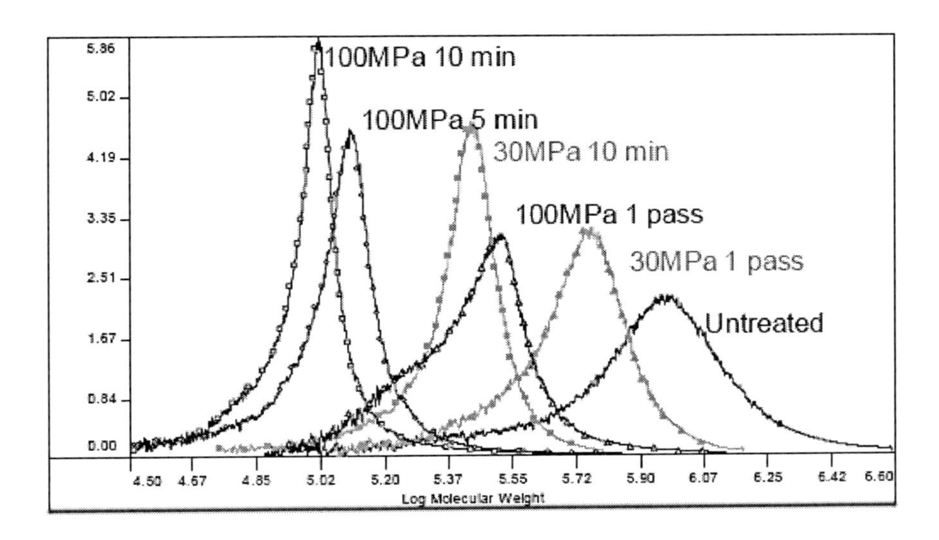

Figure 8. Effect of high pressure homogenisation (microfluidizer) on the molar mass of soluble oat beta-glucan analysed by HPSEC-RI-DALS-Viscotek in aqueous solution (0.15%, pH = 6.5) (adapted with the permission of Elsevier from Kivelä et al., 2010).

Usually, the extent of fragmentation is related to the amount of pressure applied in the system. Also in the case of oat β-glucan, the molar mass decreased in line with the applied energy and the molar mass distribution narrowed as previously reported (Figure 8) ((Buchholz, Zahn, Kenward, Slater,, & Barron, 2004); (Kivelä, Pitkänen, Laine, Aseyev, & Sontag-Strohm, 2010)).

The elongational and turbulent stresses derived from the high flow rate of the fluid in small dimensions and a sudden pressure drop in the nozzle valve/chamber are evidently the dominant degradation mechanism during the homogenisation ((Walstra & Smulders, 1998); (Stang, Schuchmann, & H., 2001)). However, oxidation also plays a role in sited processes. The acoustic cavitation in ultrasonication is known to degrade water-soluble polysaccharides (Weiss, Kristbergsson, & Kjartansson, 2010).

Ultrasonication creates acoustic cavitation and high pressure homogenisation may create hydrodynamic cavitation, which means that bubbles with extremely high local pressures and temperatures are produced ((Stang, Schuchmann, & H., 2001), (Halliwell & Gutteridge, 2007), (Weiss, Kristbergsson, & Kjartansson, 2010)). The energy of cavitation bubbles can induce homolytic fission of water molecules to hydroxyl and hydrogen radicals and thus cause oxidation ((Halliwell & Gutteridge, 2007); (Gogate & Pandit, 2005)). Hydrodynamic cavitation in high pressure homogenisers are shown to be even more effective in oxidising small molar mass constituents than for examples acoustic cavitation of ultrasonic systems (Gogate & Pandit, 2005).

4.9. Phase Separation

Within the uncontrolled degradation of β-glucan, its incompatibility with other macromolecules leading to phase separation, is also a fundamental instability parameter. Oat β-glucan separates from other macromolecules, and the separation is molar mass dependent as shown by whey protein, caseinate and pullulan ((Kontogiorgos, Tosh, & Wood, 2009); (Lazaridou & Biliaderis, 2009)). High molar mass β-glucan ($1300x10^3$g/mol) formed more stable systems than a low molar mass β-glucan (Kontogiorgos, Tosh, & Wood, 2009).

The lowest molar mass β-glucan studied formed gels, which dominated the process and promoted the instability (Lazaridou & Biliaderis, 2009). In dilute solutions, without any visible gel formation, aggregation may also alter the solution properties by compacting the molecules (molecule aggregates) ((Grimm, Krüger, & Burchard, 1995); (Kivelä, Nyström, Salovaara, & Sontag-Strohm, 2009b)).

CONCLUSION

In food processing, several degradation mechanisms including thermal degradation, oxidation, mechanical energy derived-degradation, aggregation and association/dissociation with other molecules work together on β-glucan. This review highlighted the degradation of β-glucan during aqueous processing, and introduced degradation mechanisms affecting oat β-glucan in aqueous environments. The review indicates the need for more kinetic and mechanism research on the acid hydrolysis, oxidation reactions and thermal degradation of β-glucan in order to manage the molecular properties during processing. Among others, the effect of co-extracted compounds and the role of aggregation are highly important parameters to include in these studies.

REFERENCES

AFSSA, A. F. (2008). Opinion on an application for assessment of the scientific basis of a claim relating to the effect on bloodstream cholesterol of soluble oat fibres consumed as part of an adjusted diet. *Referral No. 2007-SWA-0168. Available from www.affsa.fr* .

Aggarwal, P., Dollimore, D., & Heon, K. (1997). Comparative thermal analysis study of two biopolymers, starch and cellulose. *Journal of Thermal Analysis , 50*, 7-17.

Andersen, M. L., Outtrup, H., & Skibsted, L. H. (2000). Potential antioxidants in beer assessed by ESR spin trapping. *Journal of Agricultural and Food Chemistry , 48*, 3106.

Andersen, M.;Outtrup, H.;& Skibsted, L. (2000). Potential antioxidants in beer assessed by ESR spin trapping. *Journal of Agricultural and Food Chemistry , 48*, 3106.

Andersson, A. A., Armö, E., Grangeon, E., Fredriksson, H., Andersson, R., & Åman, P. (2004). Molecular weight and structure units of (1→3, 1→4)-β-glucans in dough and bread made from hull-less barley milling fractions . *Journal of cereal Science , 40*, 195-204.

Andersson, A. A., Rüegg, N., & Åman, P. (2008). Molecular weight distribution and content of water-extractable β-glucan in rye crisp bread. *Journal of Cereal Science , 47*, 399-406.

Arts, S., Mombarg, E., van Bekkum, H., & Sheldon, R. (1997). Hydrogen peroxide and oxygen in catalytic oxidation of carbohydrates and related compounds. *Synthesis , 6*, 597-613.

Asp, N.-G., & Trossing, M. (2001). The Swedish Code on health-related claims in action extended to product-specific. *Scandinavian Journal of Nutrition , 45*, 189.

Asp, N.-G., Mattsson, B., & Önning, G. (1992). Variation in dietary fibre, β-glucan, starch, protein, fat and hull content of oats grown in Sweden 1987-1989. *European Journal of Clinical Nutrition , 46*, 31-37.

Autio, K., Myllymäki, O., & and Mälkki, Y. (1987). Flow properties of solutions of oat β-glucans. Journal of Food Science. *52*, 1364-1366.

Bae, I. Y., Kim, S. M., & Lee, H. G. (2010). Effect of enzymatic hydrolysis on cholesterol-lowering activity of oat β-glucan. *New Biotechnology , 27*, 85-88.

Bae, I. Y., Lee, S., Kim, S. M., & Lee, H. G. (2009). Effect of partially hydrolyzed oat β-glucan on the weight gain and lipid profile of mice. *Food Hydrocolloids , 23*, 2016-2021.

Beck, E.;Tosh, S.;Batterham, M.;Tapsell, L.;& Huang, X. (2009). Oat beta-glucan increases postprandial cholecystokinin levels, decreases insulin response and extends subjective satiety in overweight subjects. *Molecular nutrition Food Research , 53*, 1343-1351.

Beer, M. U., Wood, P. J., Weisz, J., & Fillion, N. (1997). Effect of cooking and storage on the amount and molecular weight of (1→3)(1→4)- β-D-glucan extracted from oat products by an in vitro digestion system. *Cereal Chemistry , 74*, 705-709.

Beer, M., Arrigoni, E., & Amadò, R. (1995). Effects of oat gum on blood cholesterol levels in healthy young men. *European Journal of Clinical Nutrition , 49*, 517-522.

BeMiller, J. (2007). Polysaccharides: Occurrence, Structures, and Chemistry. Teoksessa J. BeMiller (Toim.), *Carbohydrate chemistry for food scientists* (2nd edition p., ss. 93-117). St.Paul, MN: American Association of Cereal Chemists.

Biörklund, M., van Rees, A., Mensink, R., & Önning, G. (2005). Changes in serum lipids and postprandial glucose and insulin concentrations after consumption of beverages with b-glucans from oats or barley:a randomised dose-controlled trial. *European Journal of Clinical Nutrition , 59*, 1272–1281.

Bradley, T. D., & Mitchell, J. R. (1988). The determination of the kinetics of polysaccharide thermal degradation using high temperature viscosity measurements. *Carbohydrate Polymers , 9*, 257-267.

Brummer, Y.;Jones, S.;Tosh, S. M.;& Wood, P. J. (2008). Extraction and physicochemical characterization of rye β-glucan and effects of barium on polysaccharide molecular weight. *Cereal Chemistry , 85*, 174-181.

Buchholz, B., Zahn, J., Kenward, M., S. G., & Barron, A. (2004). Flow-induced chain scission as a physical route to narrowly distributed, high molar mass polymers. *Polymer , 45*, 1223-1234.

Buettner, G., & Jurkiewicz, B. (1996). Catalytic metals, ascorbate and free radicals: combinations to avoid. *Radiation Research , 145*, 532-541.

Buliga, G. S., Brant, D. A., & Fincher, G. B. (1986). The sequence statistics and solution conformation of a barley (1→3)(1→4)-β-D-glucan. , 1(157),. *Carbohydrate Research , 157*, 139-156.

Burchard, W. (1995). *Light Scattering techniques.* (S. Ross-Murphy, Ed.) London: Blackie Academic and Professional, Kings College.

Burkus, Z., & Temelli, F. (2003). Determination of the molecular weight of barley beta-glucan using intrinsic viscosity measurements. *Carbohydrate Polymers , 54*, 51-57.

Burkus, Z., & Temelli, F. (1998). Effect of extraction conditions on yield, composition, and viscosity stability of barley β-Glucan gum. *Cereal Chemistry , 75*, 805-809.

Böhm, N., & Kulicke, W. (1999a). Rheological studies of barley (1→3)(1→4)- β-D-glucan in concentrated solution: investigation of theviscoelastic flow behaviour in the sol state. *Carbohydrate Research , 315*, 293–301.

Böhm, N., & Kulicke, W. (1999b). Rheological studies of barley (1→3)(1→4)-β-glucan in concentrated solution: Mechanistic and kinetic investigation of the gel formation. *Carbohydrate Research , 315*, 302-311.

Chevion, M. (1988). A site-specific mechanism for free radical induced biological damage: the essential role of redox-active transition metals. *Free Radical Biology , 5*, 27-37.

Cleary, L. J., Andersson, R., & Brennan, C. S. (2007). The behaviour and susceptibility to degradation of high and low molecular weight barley β-glucan in wheat bread during baking and in vitro digestion. *Food Chemistry , 102*, 889-897.

Cui, W., Wood, P. J., Blackwell, B., & Nikiforuk, J. (2000). Physicochemical properties and structural characterization by two-dimensional NMR spectroscopy of wheat β-D-glucan—comparison with other cereal β-D-glucans. *Carbohydrate Polymers , 41*, 249-258.

Davídek, T., Robert, F., Devaud, S., Arce Vera, F., & Blank, I. (2006). Sugar fragmentation in the maillard reaction cascade: formation of short-chain carboxylic acids by a new oxidative α-dicarbonyl cleavage pathway. *Journal of Agricultural and Food Chemistry , 54*, 6677-6684.

Davis, J. M., Murphy, E. A., Brown, A. S., Carmichael, M. D., Ghaffar, A., & Mayer, E. P. (2004). Effects of oat β-glucan on innate immunity and infection after exercise stress. *Medicine & Science in Sports & Exercise , 36*, 1321-1327.

Dawkins, N. L., & Nnanna, I. A. (1995). Studies on oat gum [(1→3, 1→4)-β-D-glucan]: Composition, molecular weight estimation and rheological properties. *Food Hydrocolloids , 9*, 1-7.

Degutyte-Fomins, L., Sontag-Strohm, T., & Salovaara, H. (2002). Oat bran fermentation by rye sourdough. *Cereal Chemistry , 79*, 345-348.

Doublier, J.-L., & Wood, P. J. (1995). Rheological properties of aqueous solutions of (1→3)(1→4)-β-D-glucan from oats (Avena sativa L.). *Cereal Chemistry , 72*, 335-340.

EFSA. (2009). Scientific Opinion on the substantiation of health claims related to beta-glucans and maintenance of normal blood cholesterol concentrations and maintenance or achievement of a normal body weight pursuant to Article 13(1) of Regulation (EC) No1924/2006. *EFSA Journal*, *7*, 1254.

FDA. (1997). FDA, 21 CFR Part 101. Food labeling, health claims: soluble dietary fiber from certain foods and coronary heart disease. *Federal Register*, *62*, 3584–3601.

Fincher, G. B. (2009). Exploring the evolution of (1→3)(1→4)-β-D-glucans in plant cell walls: Comparative genomics can help! . *Current opinion in plant biology*, *12*, 140-147.

Floury, J., Desrumaux, A., Axelos, M., & Legrand, J. (2002). Degradation of methylcellulose during ultra-high pressure homogenisation. *Food Hydrocolloids*, *16*, 47-53.

Frank, J., Sundberg, B., Kamal-Eldin, A., Vessby, B., & Åman, P. (2004). Yeast-leavened oat breads with high or low molecular weight ß-glucan do not differ in their effects on blood concentrations of lipids, insulin, or glucose in humans. *The Journal of Nutrition*, *134*, 1384-1388.

Fry, S. (1998). Oxidative scission of plant cell wall polysaccharides by ascorbate-induced hydroxyl radicals. *Biochemical Journal*, *322*, 507-515.

Gilbert, B. C., King, D. M., & Thomas, C. B. (1984). The oxidation of some polysaccharides by the hydroxyl radical: An e.s.r. investigation. *Carbohydrate research*, *125*, 217-235.

Gogate, P. R., & Pandit, A. B. (2005). A review and assessment of hydrodynamic cavitation as a technology for the future. *Ultrasonics Sonochemistry*, *12*, 21-27.

Gómez, C., Navarro, A., Manzanares, P., Horta, A., & Carbonell, J. V. (1997a). Physical and structural properties of barley (1→3),(1→4)-β-D-glucan. Part I. Determination of molecular weight and macromolecular radius by light scattering. *Carbohydrate Polymers*, *32*, 7-15.

Gómez, C., Navarro, A., Manzanares, P., Horta, A., & Carbonell, J. V. (1997b). Physical and structural properties of barley (1→3),(1→4)-β-D-glucan. Part II. Viscosity, chain stiffness and macromolecular dimensions. *Carbohydrate Polymers*, *32*, 17-22.

Graf, E., Empson, K. L., & Eaton, J. W. (1987). Phytic acid. A natural antioxidant. *The Journal of Biological Chemistry*, *262*, 11647-11650.

Grimm, A., Krüger, E., & Burchard, W. (1995). Solution properties of (1→3)(1→4)-β-D -glucan isolated from beer. *Carbohydrate Polymers*, *27*, 205-214.

Guillon, F., & Champ, M. (2000). Structural and physical properties of dietary fibres, and consequences of processing on human physiology. *Food Research International , 33*, 233-245.

Halliwell, B., & Gutteridge, J. M. (2007). *Free radicals in biology and medicine* (4th ed.). Oxford: Oxford University Press.

Hill, S. E., & Gray, D. A. (1999). Effect of sulphite and propyl gallate or ferulic acid on the thermal depolymerisation of food polysaccharides. *Journal of the Science of Food and Agriculture , 79*, 471-475.

Hjerde, T., Kristiansen, T. S., Stokke, B. T., Smidsrod, O., & Christensen, B. E. (1994). Conformation dependent depolymerisation kinetics of polysaccharides studied by viscosity measurements. *Carbohydrate Polymers , 24*, 265-275.

Hjerde, T., Smidsrød, O., Stokke, B., E., B., & Christensen, B. (1998). Acid hydrolysis of κ- and ι-carrageenan in the disordered and ordered conformations: Characterization of partially hydrolyzed samples and single-stranded oligomers released from the ordered structures. *Macromolecules , 31*, 1842-1851.

Immerstrand, T. K., Wange, C., Rascon, A., Hellstrand, P., & Nyman, M. (2010). Effects of oat bran, processed to different molecular weights of β-glucan, on plasma lipids and caecal formation of SCFA in mice. *British Journal of Nutrition , 104*, 364-374.

Izydorczyk, M. S., & Biliaderis, C. G. (2000). Structural and functional aspects of cereal arabinoxylans and b-glucans. In G. G. Doxastakis, & V. Kiosseoglou (Eds.), *Novel macromolecules in food systems* (pp. 361-384). Amsterdam: Elsevier.

JHCI, J. H. (2006). Generic Claims - oats and reduction of blood cholesterol. *Available from http://www.jhci.org.uk/approv/oats.htm* .

Johansson, L., Tuomainen, P., Anttila, H., Rita, H., & Virkki, L. (2007). Effects of processing on the extractability of oat β-glucan. *Food Chemistry , 105*, 1439-1445.

Johansson, L., Virkki, L., Anttila, H., Esselström, H., Tuomainen, P., & Sontag-Strohm, T. (2006). Hydrolysis of β-glucan. *Food Chemistry , 97*, 71-79.

Joint Health Claims Initiative, J. (2006). Generic Claims - oats and reduction of blood cholesterol. *Available from http://www.jhci.org.uk/ approv/oats.htm* .

Kaneda, H., Kano, Y., Osawa, T., Ramarathnam, N., Kawakishi, S., & Kamada, K. (1988). Detection of free radicals in beer oxidation. *Journal of Food Science , 53*, 885-888.

Kasaai, M. R., Charlet, G., Paquin, P., & Arul, J. (2003). Fragmentation of chitosan by microfluidization process. *Innovative Food Science and Emerging Technologies* , *4*, 403-413.

Keenan, J., Pins, J., Frazel, C., Moran, A., & Turnquist, L. (2002). Oat ingestion reduces systolic and diastolic blood pressure in patients with mild or borderline hypertension: A pilot trial. *The Journal of Family Practise* , *51*, 369.

Kerckhoffs, D., Hornstra, G., & Mensink, R. P. (2003). Cholesterol-lowering effect of beta-glucan from oat bran in mildly hypercholesterolemic subjects may decrease when beta-glucan is incorporated into bread and cookies. *The American Journal of Clinical Nutrition* , *78*, 221-227.

Kivelä, R., Gates, F., & Sontag-Strohm, T. (2009a). Degradation of cereal beta-glucan by ascorbic acid induced oxygen radicals. *Journal of Cereal Science* , *49*, 1-3.

Kivelä, R., Henniges, U., Sontag-Strohm, T., & Potthast, A. (2011b). Molecular changes and carbonyl group formation in the beta-glucan chain in aqueous processes.

Kivelä, R., Nyström, L., & Sontag-Strohm, T. (2008). Gel structure protects cereal β-glucan from radical induced degradation in aqueous systems. *Annual Transactions of the Nordic Rheology Society* , *16*, 185-188.

Kivelä, R., Nyström, L., Salovaara, H., & Sontag-Strohm, T. (2009b). Role of oxidative cleavage and acid hydrolysis of oat beta-glucan in modelled beverage conditions. *Journal of Cereal Science* , *50*, 190-197.

Kivelä, R., Pitkänen, L., Laine, P., Aseyev, V., & Sontag-Strohm, T. (2010). Influence of homogenisation on the solution properties of oat β-glucan . *Food Hydrocolloids* , *24*, 611-618.

Kivelä, R., Sontag-Strohm, T., Loponen, J., Tuomainen, P., & Nyström, L. (2011a). Oxidative and radical mediated cleavage of beta-glucan in thermal treatments. *Carbohydrate Polymers* .

Knill, C. J., & Kennedy, J. F. (2003). Degradation of cellulose under alkaline conditions. *Carbohydrate Polymers* , *51*, 281-300.

Kontogiorgos, V., Tosh, S. M., & Wood, P. J. (2009). Phase behaviour of high molecular weight oat β-glucan/whey protein isolate binary mixtures. *Food Hydrocolloids* , *23*, 949-956.

Krochta, J. M., Tillin, S. J., & Hudson, J. S. (1987). Degradation of polysaccharides in alkaline solution to organic acids: Product characterization and identification. *Journal of Applied Polymer Science* , *33*, 1413-1425.

Kök, M. S., Hill, S. E., & Mitchell, J. R. (1999). Viscosity of galactomannans during high temperature processing: Influence of degradation and solubilisation . *Food Hydrocolloids , 13*, 535-542.

Lagoueyte, N., & Paquin, P. (1998). Effects of microfluidization on the functional properties of xanthan gum. *Food Hydrocolloids , 12*, 365-371.

Lai, V. M., Lii, C. Y., Hung, W. L., & Lu, T. J. (2000). Kinetic compensation effect in depolymerisation of food polysaccharides. *Food Chemistry , 68*, 319-322.

Lambo, A. M., Öste, R., & Nyman, M. E. (2005). Dietary fibre in fermented oat and barley β-glucan rich concentrates. *Food Chemistry , 89*, 283-293.

Lazaridou, A., & Biliaderis, C. G. (2009). Concurrent phase separation and gelation in mixed oat β-glucans/sodium caseinate and oat β-glucans/pullulan aqueous dispersions. *Food Hydrocolloids , 23*, 886-895.

Lazaridou, A., & Biliaderis, C. G. (2007). Cryogelation phenomena in mixed skim milk powder – barley β-glucan–polyol aqueous dispersions. *Food Research International , 40*, 793-802.

Lazaridou, A., Biliaderis, C. G., & Izydorczyk, M. S. (2003). Molecular size effects on rheological properties of oat beta-glucans in solution and gels. *Food hydrocolloids , 17*, 693-712.

Lazaridou, A., Biliaderis, C., & Izydorczyk, M. (2007). *Cereal β-glucans: structure, physical properties, and physiological functions* (1st edition ed.). (C. Biliaderis, & M. Izydorczyk, Eds.) LLC: TaylorandFrancis group.

Lazaridou, A., Biliaderis, C., & Izydorczyk, M. (2003b). Molecular size effects on rheological properties of oat b-glucans in solution and gels. *Food Hydrocolloids , 17*, 693–712.

Li, W., Cui, S. W., Wang, Q., & Yada, R. Y. (2010). Studies of aggregation behaviours of cereal β-glucans in dilute aqueous solutions by light scattering: Part I. structure effects. *Food Hydrocolloids , 25*, 189-195.

Li, W., Wang, Q., Cui, S. W., Huang, X., & Kakuda, Y. (2006). Elimination of aggregates of (1→3) (1→4)-β-D-glucan in dilute solutions for light scattering and size exclusion chromatography study. *Food Hydrocolloids , 20*, 361-368.

Manthey, F. A., Hareland, G. A., & Huseby, D. J. (1999). Soluble and insoluble dietary fiber content and composition in oat. *Cereal Chemistry , 76*, 417-420.

Miller, J. G., & Fry, S. C. (2001). Characteristics of xyloglucan after attack by hydroxyl radicals. *Carbohydrate Research , 332*, 389-403.

Mitchell, J. R., Reed, J., Hill, S. E., & Rogers, E. (1991). Systems to prevent loss of functionality on heat treatment of galactomannans. *Food Hydrocolloids* , *5*, 141-143.

Moriartey, S., Temelli, F., & Vasanthan, T. (2010). Effect of formulation and processing treatments on viscosity and solubility of extractable barley β-glucan in bread dough evaluated under in vitro conditions. *Cereal Chemistry* , *87*, 65-72.

Mäkeläinen, H., Anttila, H., Sihvonen, J., Hietanen, R.-M., Tahvonen, R., Salminen, E., et al. (2007). The effect of ß-glucan on the glycemic and insulin index. *European Journal of Clinical Nutrition* , *6*, 779-785.

Mälkki, Y. (1992). Oat bran concentrates: Physical properties of β-glucan and hypocholesterolemic effects in rats. *Cereal Chemistry* , *69*, 647-653.

Naumann, E., van Rees, A. B., Önning, G., Öste, R., Wydra, M., & Mensink, R. P. (2006). Beta-glucan incorporated into a fruit drink effectively lowers serum LDL-cholesterol concentrations. *The American Journal of Clinical Nutrition* , *83*, 601-605.

Nevell, T. (1985). *Degradation of cellulose by acids, alkalis, and mechanical means.* (T. Nevell, & S. S. H. Zeronian, Eds.) Chichester: Ellis Horwood.

Nilsson, L., Leeman, M., Wahlund, K.-G., & Bergenståhl, B. (2006). Mechanical degradation and changes in conformation of hydrophobically modified starch. *Biomacromolecules* , *7*, 2671-2679.

Parikka, K.;Leppänen, A.-S.;Pitkänen, L.;Reunanen, M.;Willför, S.;& Tenkanen, M. (2010). Oxidation of polysaccharides by galactose oxidase. *Journal of Agricultural and Food Chemistry* , *58*, 262-271.

Park, S., Bae, I., Lee, S., & Lee, H. (2009). Physicochemical and Hypocholesterolemic Characterization of Oxidized Oat β-Glucan. *Journal of Agricultural and Food Chemistry* , *57*, 439-443.

Parrish, F., Perlin, A., & Reese, E. (1960). Selective enzymolysis of poly-β-D-glucans, and the structure of the polymers. *Canadian Journal of Chemistry* , *38*, 2094-2104.

Paterson, L. A., Hill, S. E., Mitchell, J. R., & Blanshard, J. M. (1997). Sulphite and oxidative—reductive depolymerization reactions . *Food Chemistry* , *60*, 143-147.

Peniche-Covas, C., Argüelles-Monal, W., & San Román, J. (1993). A kinetic study of the thermal degradation of chitosan and a mercaptan derivative of chitosan. *Polymer Degradation and Stability* , *39*, 21-28.

Picout, D. R., & Ross-Murphy, S. B. (2007). On the Mark–Houwink parameters for galactomannans. *Carbohydrate Polymers* , *70*, 145-148.

Pielichowski, K., & Njuguna, J. (2005). *Thermal degradation of polymeric materials* (1st ed.). Shawbury: Rapra Technology Limited.

Platt, S., & Clydesdale, F. (1984). Binding of iron by cellulose, lignin, sodium phytate and beta-glucan, alone and in ombination, under simulated gastrointestinal pH conditions. *Journal of Food Science , 49*, 531-535.

Potthast, A., Kostic, M., Schiehser, S., Kosma, P., & Rosenau, T. (2007). Studies on oxidative modifications of cellulose in the periodate system: Molecular weight distribution and carbonyl group profiles. *Holzforschung , 61*, 662-667.

Potthast, A., Rosenau, T., & Kosma, P. (2006). *Analysis of Oxidized Functionalities in Cellulose. Polysaccharides II.* (1st ed.). (D. Klemm, Ed.) Berlin, Heidelberg: Springer.

Qian, S. Y., & Buettner, G. R. (1999). Iron and dioxygen chemistry is an important route to initiation of biological free radical oxidations: An electron paramagnetic resonance spin trapping study. *Free Radical Biology and Medicine , 26*, 1447-1456.

Regand, A., Tosh, S., Wolever, T. ..., & Wood, P. (2009). Physicochemical properties of β-Glucan in differently processed oat foods influence glycemic response. *Journal of Agricultural and Food Chemistry , 57*, 8831-8838.

Ren, Y., Ellis, P. R., Ross-Murphy, S. B., Wang, Q., & Wood, P. J. (2003). Dilute and semi-dilute solution properties of (1-> 3),(1-> 4)-beta-d-glucan, the endosperm cell wall polysaccharide of oats (avena sativa L.). *Carbohydrate Polymers , 53*, 401-408.

Rimsten, L., Stenberg, T., Andersson, R., Andersson, A., & Åman, P. (2003). Determination of β-glucan molecular weight using SEC with calcofluor detection in cereal extracts. *Cereal Chemistry , 80*, 485-490.

Ripsin, C., Keenan, J., Jacobs, D., Elmer, P., Welch, R., Van Horn, L., et al. (1992). Oat products and lipid lowering: a meta-analysis . *Journal of American Medical Association , 267*, 3317-3325.

Robert, R., Barbati, S., Ricq, N., & Ambrosio, M. (2002). Intermediates in wet oxidation of cellulose: Identification of hydroxyl radical and characterization of hydrogen peroxide. *Water Research , 36*, 4821-4829.

Ruxton, C. H., & Derbyshire, E. (2008). A systematic review of the association between cardiovascular risk factors and regular consumption of oats. *British Food Journal , 110*, 1119-1133.

SFOPH. (2006). Swiss Federal Office of Public Health. *Available from http://www.bag.admin.ch/index.html?lang=en* .

Silvestri, S., & Gabrielson, G. (1991). Degradation of tragacanth by high shear and turbulent forces during microfluidization. *International Journal of Pharmaceutics , 73*, 163-167.

Skendi, A., Biliaderis, C. G., & Lazaridou, A. (2003). Structure and rheological properties of water soluble beta-glucans from oat cultivars of avena sativa and avena bysantina. *Journal of Cereal Science , 38*, 15-31.

SkåneDairy. (2002). Primaliv first food product in Sweden to be approved and labelled as a functional food.extract from press release from Sweden approves its first health claim. *at http://www.creanutrition-sof.com/fileadmin/template/creanutrition/files_redakteur/news_internet/NNB_ Sweden_Primaliv_claim_Sep_2002_.pdf* .

SNF. (2001). Health claims in the labeling and marketing of food products. The food sector's code of practice. 18-19. *Available from http://www.hp-info.nu/SweCode_2004_1.pdf* .

Soldi, W. (2005). *Stability and degradation of polysaccharides.* (S. Dimitriu, Ed.) New York: Marcel Dekker.

Stang, M., Schuchmann, H., & H., S. (2001). Emulsification in high-pressure homogenizers. *Engineering in Life Sciences , 1*, 151-157.

Story, J. A., & Kritchevsky, D. (1976). Comparison of the binding of various bile acids and bile salts in vitro by several types of fiber. *Journal of Nutrition , 106*, 1292-1294.

Strlic, M., Kocar, D., Kolar, J., Rychlý, J., & Pihlar, B. (2003). Degradation of pullulans of narrow molecular weight distribution—the role of aldehydes in the oxidation of polysaccharides. *Carbohydrate Polymers , 54*, 221-228.

Symons, L. J., & Brennan, C. S. (2004). The effect of barley beta-glucan fiber fractions on starch gelatinization and pasting characteristics. *Journal of Food Science , 69*, 257-261.

Temelli, F. (1997).). Extraction and functional properties of barley beta-glucan as affected by temperature and pH. *Journal of Food Science , 62*, 1194-1201.

Tosh, S. M., Brummer, Y., Wolever, T. M., & Wood, P. J. (2008). Glycemic response to oat bran muffins treated to vary molecular weight of β-glucan. *Cereal Chemistry , 85*, 211-217.

Tosh, S. M., Wood, P. J., Wang, Q., & Weisz, J. (2004). Structural characteristics and rheological properties of partially hydrolyzed oat β-glucan: The effects of molecular weight and hydrolysis method. *Carbohydrate Polymers , 55*, 425-436.

Tosh, S., Brummer, Y., Miller, S., Regand, A., Defelice, C., Duss, R., et al. (2010). Processing affects the physicochemical properties of β-glucan in oat bran cereal. *Journal of Agricultural and Food Chemistry , 58*, 7723-7730.

Törrönen, R., Kansanen, L., Uusitupa, M., Hänninen, O., Myllymäki, O., Härkönen, H., et al. (1992). Effects of oat bran concentrate on serum lipids in free-living men with mild to moderate hypercholesterolaemia. *European Journal of Clinical Nutrition , 46*, 621-627.

Vahouny, G. V., Tombes, R., Cassady, M. M., Kritchevsky, D., & Gallo, L. (1980). Dietary fibers: V. Binding of bile salts, phospholipids and cholesterol from mixed micelles by bile acid sequestrants and dietary fibers. *Lipids , 15*, 1012-1018.

Vaikousi, H., & Biliaderis, C. G. (2005). Processing and formulation effects on rheological behavior of barley β-glucan aqueous dispersions . *Food Chemistry , 91*, 505-516.

Valko, M., Morris, H., & Cronin, M. T. (2005). Metals, toxicity and oxidative stress. *Current Medicinal Chemistry , 12*, 1161-1208.

Vallès-Pàmies, B., Barclay, F., Hill, S. E., Mitchell, J. R., Paterson, L. A., & Blanshard, J. M. (1997). The effects of low molecular weight additives on the viscosities of cassava starch . *Carbohydrate Polymers , 34*, 31-38.

Walstra, P., & Smulders, P. (1998). *Emulsion formation.* (B. Binks, Ed.) Cambridge: Royal Society of Chemistry.

Wang, Q., Ellis, P. R., & Ross-Murphy, S. B. (2000). The stability of guar gum in an aqueous system under acidic conditions. *Food Hydrocolloids , 14*, 129-134.

Wang, Q., Wood, P. J., & Ross-Murphy, S. B. (2001). The effect of autoclaving on the dispersibility and stability of three neutral polysaccharides in dilute aqueous solutions. *Carbohydrate Polymers , 45*, 355-362.

Wardman, P., & Candeias, L. P. (1996). Fenton chemistry: An introduction. *Radiation Research , 145*, 523-531.

Weiss, J., Kristbergsson, K., & Kjartansson, G. (2010). *Engineering food ingredients with high-intensity ultrasound.* (1st ed.). (F. F.Hao, J. Weiss, & G. Barbosa-Canovas, Eds.) New York: Springer.

Welch, K. D., Davis, T. Z., & Aust, S. D. (2002). Iron autoxidation and free radical generation: Effects of buffers, ligands, and chelators. *Archives of Biochemistry and Biophysics , 397*, 360-369.

Whistler, R. L., & BeMiller, J. N. (1958). Alkaline degradation of polysaccharides. *Advances in Carbohydrate Chemistry , 13*, 289-329.

Voedingscentrum. (2005). Assessment Report, 19-04-05, on Pró-FIT® bread, containing 2.2 g β-glucan from OatWell® oat bran per 100 g bread. *Netherlands* .

Wolever, T., Tosh, S., Gibbs, A., Brand-Miller, J., Duncan, A., Hart, V., et al. (2010). Properties of oat β-glucan influence its LDL cholesterol lowering effect in humans. *FASEB Journal , 24*, 723-732.

von Sonntag, C. (1980). *Free-radical reactions of carbohydrates as studied by radiation techniques.* (R. Tipson, & D. D. Horton, Eds.) New York: Academic Press.

Wood, P. (1993). *Current practise and novel processes* (1st ed.). (P. Wood, Ed.) St.Paul, MN: American Association of Cereal Chemists .

Wood, P. J. (2010). Oat and rye β-glucan: Properties and function. *Cereal Chemistry , 87*, 315-330.

Wood, P. J., Beer, M. U., & Butler, G. (2000). Evaluation of role of concentration and molecular weight of oat beta-glucan in determining effect of viscosity on plasma glucose and insulin following an oral glucose load. *The British Journal of Nutrition , 84*, 19-23.

Wood, P. J., Weisz, J., Fedec, P., & Burrows, V. B. (1989). Large-scale preparation and properties of oat fractions enriched in (1-43)(1-4)-β-D-glucan. *Cereal Chemistry , 66*, 97-103.

Wood, P. (1991). Oat β-glucan-physicochemical properties and physiological effects. *Trends in Food Science and Technology , 2*, 311-314.

Wu, J., Zhang, Y., Wang, L., Xie, B., Wang, H., & Deng, S. (2006). Visualization of single and aggregated hulless oat (Avena nuda L.) (1→3),(1→4)-β-d-glucan molecules by atomic force microscopy and confocal scanning laser microscopy. *Journal of Agricultural and Food Chemistry , 54*, 925-934.

Vårum, K. M., Smidsrod, O., & Brant, D. A. (1992). Light scattering reveals micelle-like aggregation in the (1→3),(1→4)-β-D-glucans from oat aleurone. *Food Hydrocolloids , 5*, 497-511.

Zoldners, J., Kiseleva, T., & Kaiminsh, I. (2005). Influence of ascorbic acid on the stability of chitosan solutions. *Carbohydrate Polymers , 60*, 215-218.

Åman, P., Rimsten, L., & Andersson, R. (2004). Molecular weight distribution of β-glucan in oat-based foods. *Cereal Chemistry , 81*, 356-360.

Önning, G., Wallmark, A., Persson, M., Åkesson, B., Elmståhl, S., & Öste, R. (1999). Consumption of oat milk for 5 weeks lowers serum cholesterol and LDL cholesterol in free-living men with moderate hypercholesterolemia. *Annals of Nutrition and Metabolism , 43*, 301-309.

In: Oats: Cultivation, Uses and Health Effects ISBN 978-1-61324-277-3
Editor: D. L. Murphy, pp. 97-123

Chapter 3

An Overview of Oat Contamination with Mycotoxins: Strategies for Their Control

Angel Medina and Naresh Magan
Applied Mycology Group, Cranfield Health, Vincent Building, Cranfield University, Cranfield, Bedfordshire. MK43 0AL, UK

Abstract

Oats, like other temperate cereals, can be invaded by a range of fungi both in the field before harvest and during storage. Some of these fungal species, belonging mainly to the genera *Fusarium*, *Aspergillus* and *Penicillium*, can produce a wide range of secondary metabolites. Some have the ability to contaminate the ripening grain pre-harvest or grain post-harvest with mycotoxins which can be toxic to animals and humans. Some are proven or suspected carcinogens.

Most mycotoxins are very heat stable so that once formed they are difficult to eliminate from the food supply. Cereals used for animal feed often utilise co-products that in general are likely to contain higher amounts of mycotoxins. The animal tissues will take up these chemical compounds and in this way the mycotoxins can be re-introduced into the human food chain.

Because of this, from the mid 1980's onwards,and due to the use of oat for human nutrition becoming more popular, many studies have described contamination of oats with different mycotoxin families including fusariotoxins (nivalenol, deoxynivalenol, zearalenone, HT-2 and T-2), ochratoxins and aflatoxins.

This chapter will be focused on the contamination of oats with different mycotoxins with special attention given to those produced by *Fusarium* species, the ecology of the species involved, and the most recent data on levels of contamination with Type A trichotecenes in northern European countries. The European legislation regarding mycotoxin contamination of cereals will be highlighted and, finally, reduction and control strategies will be discussed.

1. INTRODUCTION

The term mycotoxin is derived from the Greek word "Mykes" meaning mold and the Latin word "Toxicum" meaning toxic. Many staple crops, and among them oats, can be invaded by a range of fungi both in the field before harvest and during storage. Some of these fungal species, belonging mainly to the genera *Fusarium*, *Aspergillus* and *Penicillium*, can produce a wide number of secondary metabolites. Among them some have the ability to contaminate the ripening grain pre-harvest, during grain drying and storage with mycotoxins which can be toxic to animals and humans. It is estimated that 10-30% of the harvested grain is lost due to fungal spoilage and the Food and Agriculture Organisation (FAO) of the United Nations estimates that about 25% of the world food crops are affected by mycotoxins.

Belonging to different families, mycotoxins can be present in a wide range of food products (grain crops, some fruits, vegetables and herbs, and some nuts and legumes).Cereals are important targets as they are staple foods, and nutritionally conducive to colonisation and are stored for the medium and long term in large quantities.

These can have a significant effect on human and animal health because they can be carcinogenic (e.g. aflatoxins) and mycotoxins are very heat-stable and thus difficult to destroy during processing. Moreover, cereals used for animal feed often utilise co-products that in general are likely to contain higher amounts of mycotoxins. The animal tissues will take up these chemical compounds and they can thus be re-introduced into the human food chain. This has resulted in strict legislative limits in many parts of the world for mycotoxins in a wide range of foodstuffs (European Commission, 2006).

More than 500 metabolites have been described, but among them, the most economically important from the human health point of view are the aflatoxins (AFL), ochratoxin A (OTA), fumonisins (FUM), trichothecenes (TCT) and zearalenone (ZEA). Table 1 summarizes the most important mycotoxins, the effects they can have and the more relevant mycotoxigenic species.

From the mid 1980's the use of oats for human nutrition became more popular and many authors were interested in the analysis of the mycotoxin content of oats. Oats were reported as a functional food as it contains soluble fibres (β-glucan) that have been shown to reducecholesterol promoting healthy lifestyle. As an example, if we examine the UK situation, the use of oats for human food production has been increasing gradually in the past 25 years and trendssuggest that this willcontinue in the future, possibly due to its enhanced image as a healthy life-style food. Around 40% of the UK oat production is used for the production of oat-based foods. The remaining grains are used in animal feed, as seed stock or exported.

Table 1.Summary of the major mycotoxins and effects that they cause with the associated species

Toxins / Effects	AFL	OTA	FUM	TCT	ZEA
Carcinogenic	X	X	X		
Hepatotoxic	X	X	X		
Immunotoxic	X	X	X	X	
Nephrotoxic		X	X		X
Neurotoxic			X	X	
Oestrogenic					X
Teratogenic	X	X			X
Fungi	*Asper-gillus flavus,A. parasiticus, A. nomius*	*A. carbo-narius, A. wester-dijkiae, Penicillium verrucosum*	*Fusarium verti-cillioides, F. prolife-ratum.*	*Fusarium graminearum, F. culmorum, F. langsethiae, F. sporo-trichioides*	*F. culmo-rum, F. poae*

Contamination of oats with different mycotoxin families includes fusariotoxins (nivalenol, deoxynivalenol, zearalenone, HT-2 and T-2), ochratoxins and aflatoxins. These have all been found in different regions of the world.In this chapter we will focus on the contamination of oats with different mycotoxins with special attention given to those produced by *Fusarium* species.

2. *FUSARIUM* MYCOTOXINS

Fusarium species require somewhat lower temperatures for growth and mycotoxin production than the other mycotoxigenic species as Aspergillus, so that, the fusariotoxins have traditionally been associated with temperate cereals. However, the available data indicates the global scale of contamination of cereal grains with a number of *Fusarium* mycotoxins (Muller and Schwadorf, 1993; Chulze et al., 1996; Viquez et al., 1996, BIOMIN's Mycotoxin Survey, 2010).

Fusarium species synthesise a wide range of mycotoxins of diverse structure and chemistry (Flannigan, 1991).The most important from the point of view of animal health and productivity are the trichothecenes, zearalenone, and the fumonisins (D'Mello et al., 1997).

2.1. Type B Trichothecenes and ZEA in Oats

2.1.1. Description and Toxicology

The trichothecenes are cyclic squiterpenoids and according their chemical structure are divided into four basic groups (type A-D), while types A and B are the most important members.

Type B trichothecenes include deoxynivalenol (DON, also known as vomitoxin) and its 3-acetyl and 15-acetyl derivatives (3-ADON and 15-ADON, respectively), nivalenol (NIV) and fusarenon-X. The synthesis of the two types of trichothecenes appears to be characteristic for a particular *Fusarium* species.

Toxicologically, trichothecenes produce a number of common effects including skin irritation, haemorrhagic lesions and depression of immune response. In humans, the major responses to trichothecene ingestion have been reported to be vomiting, diarrhoea and skin reaction.

DON is one of the most studied *Fusarium*trichothecenes which, when ingested in high doses by farm animals, causes vomiting and diarrhoea; at lower doses, monogastric animals like pigs exhibit feed refusal and weight loss

(Anon, 1999; Visconti, 2001; Anon, 2004). Although being the more studied and most prevalent in small-grain cereals, other trichothecenes have the same cellular activity which is disruption of protein synthesis, and have a higher cellular toxicity than DON. Nivalenol and T-2 (a Type A trichothecene) are ca. 20 times more toxic than DON, although there are relative differences dependent on the target cell or animal studied (Desjardins, 2006).

Many *Fusarium* species are able to synthesisezearalenone (ZEA). Produced late in the crop growing season, near to harvest, ZEA has no known function in the fungus (Matthauset al.2004). Zearalenone has low cellular toxicity but is problematic as it has high estrogenic activity causing hyperoestrogenism in animals and humans. In animals the mycotoxin causes a range of fertility problems, with young female pigs being particularly susceptible (Anon, 2004).Its co-occurrence with certain trichothecenes raises important issues regarding additivity and/or synergism in the aetiology of mycotoxicoses in animals.

2.1.2. Incidence

In 1986, Ueno et al. (1986) published some data regarding contamination of cereals with NIV, DON and ZEA and they found contamination with the first two toxins in one sample they analysed from USSR.

Tanaka et al. (1988) described the presence of NIV in one sample out of eight, and DON and ZEA in 3 samples out of eight, in oat samples from the southern part of West Germany. DON and ZEA were found in one sample out of two, in cereals from the northern part. In the same work, the presence of ZEA was described in one out of 5 samples of oats from Italy, and NIV and ZEA were detected with high frequency in samples from Nepal (4 and 5 out of sevenrespectively).

Subsequently, Tanaka et al. (1990) again studied the presence of fusariotoxinsbut this time in cereals harvested in The Netherlands and they found DON, NIV and ZEA. Levels were 0.056-0.147, 0.017-0.039 and 0.016-0.029 $\mu g\ g^{-1}$respectively.

Hietaniemi and Kumpulainen (1991) described the presence of DON in Finish oats at levels ranging from1.3 to 2.6 mg kg^{-1}.

In 1996, Langseth and Elen described the presence of DON and NIV in Norwegian oats produced by commercial growers. Oat grains were found to be more heavily contaminated with DON than barley or wheat kernels. Levels for DON ranged from 7.2 to 62.05 mg kg^{-1} and the maximum level of NIV they found was 0.67. This was attributed to edaphic and agronomic factors and to different infection pathways by *Fusarium* pathogens (Langseth and Elen, 1996).

In 2001 the SCOOP (SCOOP: Scientific Co-operation on Questions relating to Food) task 3.2.10 "Collection of occurrence data of *Fusarium* toxins in food and assessment of dietary intake by the population of EU Member States" was established and the results were published in 2004 (Schothorst and van Egmond, 2004). Presence of DON and NIV in European cereals was described. Approximately one third on oats samples analysed for DON were positive. In the case of NIV this percentage was lesser only being contaminated one fifth.

More recently, a survey of retail oat products was carried out by the United Kingdom Food Standard Agency between May and October 2003. Three hundred and thirty five samples of retail oat products were analysed for the presence of trichothecenes including among others DON, NIV, T-2 toxin and HT-2 toxin (Food Standards Agency, 2004). Trichothecene toxins were quantified in 184 of 335 samples although, in most cases, the levels found were low. A total of 6 trichothecenemycotoxins were detected, with DON being one of the most common.

Presence of DON and ZEA has been described in Canadian breakfast oat-based cereals by Roscoe et al. (2008). These authors analysed 27 cereals samples obtained from 1999 to 2001 and they found DON in 63% (17 positive samples) of the samples with anoverall mean of 20 ng g^{-1}.ZEA wasdetected in 11% (3/27) of the oat-based samplesat an overall mean level of 0.5 ng g^{-1}.

2.2. T-2 and HT-2 Toxin in Oats, an Actual Concern

2.2.1. Description and Toxicology

T-2 and HT-2 toxins are type-A trichothecenes produced by different *Fusarium* species (Bottalico, 1998; Torpandand Adler, 2004; TorpandNiremberg, 2004). T-2 toxin is rapidly metabolized to HT-2 toxin which is also the main metabolite in vivo (Eriksenand Alexander, 1998; Visconti, 2001). Studies on the metabolism of T-2 (Matsumoto et al., 1978) have suggested that the liver is the major organ for its metabolism, although other tissues are capable of metabolic modification of this toxin. Hepatic carboxylesterases have been shown to be responsible for the specific deacetylation of T-2, resulting in HT-2 as the major metabolite (Matsumoto et al., 1978; Johnsen et al., 1988). T-2 toxin, the most toxic Type A trichothecene, is a potent inhibitor of DNA, RNA, protein synthesis and mitochondrial function, and shows immunosuppressive and cytotoxic effects both *in vivo* and *in vitro* (Visconti et al., 1991; Canady et al., 2001; Visconti, 2001). HT-2 and T-2 were implicated in Alimentary Toxic Aluekia caused by the consumption of cereals which had overwintered in fields in Russia in the 1940s (Desjardins, 2006).

The ability to produce T-2 and HT-2 toxins is well known for different *Fusarium* species such as *F. acuminatum*, *F. sporotrichioides* and *F. poae* (Bottalico, 1998; Torpand Adler, 2004; TorpandNiremberg, 2004), but, more recently, studies in Norwegian cereal grain (Kosiak et al., 1997; Langseth and Rundberget, 1999) led to the description of *Fusariumlangsethiae* (Torp and Langseth, 1999; Torp and Nirenberg, 2004).This species is also able to produce type A trichothecenes, especially the highly toxic T-2 and HT-2 toxins, and diacetoxyscirpenol (DAS) (Thrane et al., 2004) and has been isolated from infected oats, wheat and barley in central and northern Europe (Torp and Adler, 2004; Torp and Nirenberg 2004; Hudec and Rohácik, 2009). One problem is that this species can be readily isolated from small-grain cereals from symptomless oats, barley and wheat grains.

There was practically no information regarding how ecological factors affect the growth and trichothecenesA production by*F. langsethiae*. As this species became very important in northern Europe in a range of small grains Medina and Magan (2010, 2011) reported on the effect of a_w x temperature on growth and T-2 and HT-2 production of 8 strains, two each from the U.K., Norway, Sweden and Finland on an oat-based medium (Medina and Magan, 2010, 2011).

This showed that there were no statistical differences in terms of water activity (a_w, 0.995-0.90) and temperature (10-37°C) tolerances of the strains in terms of growth. The $a_w$and temperature optima were 0.98-0.995 and 25°C and growth limits were established at 0.92-0.93 a_w and 37°C and 5°C, respectively. Regarding the toxins production, the optimum a_w and temperature conditions for T-2 + HT-2 production was between 0.98-0.995, and 20-30°C respectively. Statistical analysis showed significant differences between different a_w levels but no significant effect of the temperatures examined.Figure 1 shows the representation of growth/no growth and toxin/no toxin (T-2+HT-2) boundaries of *F. langsethiae* for different water availabilities and temperatures obtained with 8 different strains from 4 Northern Europe countries.

These new data available has shown that water availability appears to be very important in determining contamination levels with these toxins. In contrast, they are produced over a wide temperature range. Often oats are harvested late in the season when conditions are wet. Thus inefficient drying may allow *F. langsethiae* to continue to colonise and increase contamination post-harvest.

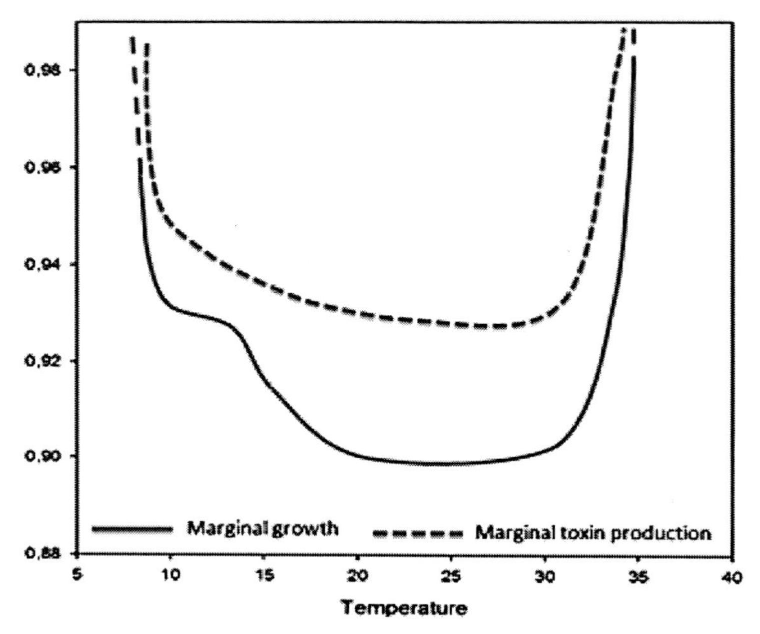

Figure 1. The representation of growth/no growth and toxin/no toxin (T-2 + HT-2) boundaries of *F. langsethiae* for different water availabilities (0.995-0.88)and temperatures (5-37°C) (Adapted from Medina et al. 2011).

2.2.2. Incidence

First information on the importance of type A trichotecenes in oats was published in the SCOOP reports (Reports on tasks for Scientific Cooperation) in 2003 (Schothorst, 2004). In this report more than 3000 cereal samples were analysed from different European countries and positive T-2 and HT-2 contamination was registered in 21 and 41% of samples respectively. Regarding oats, 464 samples were analysed for both mycotoxins. T-2 was found in 16% of samples, which ranging from LOD10 μgkg^{-1} to 550 μgkg^{-1}.The mean ranged between 4.2 to 68.3 μgkg^{-1}. HT-2 was found in 41% of positive samples, which ranged from LOD 10 $\mu g\ kg^{-1}$ to 1150 μgkg^{-1}.

Roscoe et al. (2007) measured the amount of HT-2 toxin in 27 samples of oat-based breakfast cereals and they found just one sample contaminated with 20 ng g^{-1}. Unfortunately the amount of T-2 was not studied in this work.

Mankevičienė et al. (2007) published results about the contamination on different Lithuanian cereals from the 2004 and 2005 harvests. Regàrding oats results showed that 5 samples of oats were analysed each year and in both cases 100% of positive samples for T-2 toxin were found. The overall results showed contamination levels ranging from 10.8 to 121.5 μg of T-2 kg^{-1} of oat.

The same year, Buckley et al. (2007) analysed Canadian and Irish oats. Their results showed that 40% of the 17 samples analysed were contaminated with T-2 toxin.

Gottschalk et al. (2007) studied the presence of Type A trichothecenes in oats from Germany. They introduced the agriculture practices as parameter and analysed the differences between conventional (25 samples) and organically (18 samples) produced oats. The percentage of positive samples in both agricultural practices was 100%. The mean concentration of T-2 and HT-2 (sum of the toxins) in all samples was 17±18 μg kg^{-1} (mean ± SD). Twenty-one percent and 7.1% of the samples showed T-2/HT-2 levels higher than 25 and 50 μgkg^{-1} respectively.Conventionally produced samples had a mean of T-2+HT-2 contamination of 27 ± 21 μg kg^{-1}, samples from organic production were contaminated with 7.6 ± 4.6 μg kg^{-1}. The level of HT-2 usually was about 2-fold as high as the level of T-2.

A recent study in the UK, has shown that the incidence and concentration of HT-2 and T-2 in oats was found to be high with concentrations greater than 10 ppb in 92% of samples and a combined (HT-2+T-2) median, mean and maximum concentrations of 213, 570 and 9990 ppb respectively of all analysed samples (Edwards, 2007). The incidence and concentration of HT-2 and T-2 in barley and wheat were similar, with approximately 1% of the samples exceeding 100 ppb. These results are a clear indication that these mycotoxins are more of an issue on oats rather than on wheat and barley.

Later, the same author studied the contamination in relation with different abiotic factors as year, region and practice (Edwards, 2009). Some of the most outstanding results he found were a highly significant interaction ($p<0.001$) between year andregion, with high concentrations detected in all UK regions and a highly significantdifference in the HT-2+T-2 concentration betweenorganic and conventional samples ($p<0.001$).

Organic samples were approximately five times lower,with predicted means of 50 and 264 mgkg^{-1} HT-2+T-2for organic and conventional oats, respectively.

In the previous study described here (Gottschalk 2007), also differences between conventional oatand organic oat products have been recorded, but the differencewas not as great as reported by Edwards (2009). It has been suggested that organic oats may have a lower HT-2 and T-2 contentdue to differences in a number of agronomic factors. Among the possibilities previous crop, cultivation and varietal choice seem to be implicated.Organic growers tend to have longer rotations, whichare more diverse and less cereal intensive thanconventional growers. This could result in a lowerlevel of the HT-2+T-2producing *Fusarium* inoculumwithin organic production (Edwards, 2004, 2009).

A study of Pettersson et al. (2008) in Swedish oats described much higher contamination of oats with T-2 and HT-2 in fieldtrials in 2005-2007 and especially in 2006, compared to surveys1994-1998. Authors described the co-occurrence of both toxins in oat samples and HT-2 was nearly always inhigher concentration. High portions (13-44%) of the samples were like in other NorthEuropean surveys above the suggested EU maximum level (500 $\mu g kg^{-1}$) for the sum ofT-2 and HT-2 toxins in raw oats.

As previously described in an UK study (Edward, 2007), results in Swedish trials showedthat there are varietal differences inthe fungal susceptibility and the toxin occurrence in the harvested kernels (Petterson, 2008).

3. ASPERGILLUS MYCOTOXINS

Aspergillus species are not considered to be major cause of plant disease but they are responsible for several disorders in various plant and plant products. The most common species are *A. niger* and *A. flavus*, followed by *A. parasiticus*, *A. ochraceus*, *A. carbonarius*, and *A. alliaceus*. They are able to contaminate agricultural products at different stages including pre-harvest, harvest, post-processing and handling. Changes due to spoilage by *Aspergillus* species include: pigmentation, discoloration, rotting, development of off-odours and off-flavours. Because they are opportunistic pathogens, most of them are encountered as storage moulds on plant products (Kozakiewicz 1989).

Although all this, the major healthy concern is that their presence can lead to the occurrence of mycotoxins contamination of foods and feeds. Various mycotoxins have been identified in foods and feeds contaminated by *Aspergillus* species, the most important are the aflatoxins(AF) and ochratoxin A (OTA)(Varga et al. 2004).

3.1. Aflatoxins

3.1.1. Description and Toxicology

The term "aflatoxins" normally refers to the group of difuranocoumarins and classified in two groups according to their chemical structure; the difurocoumarocyclopentenone series (Among them AFB1, AFB2, AFM1 and AFM2) and the difurocoumarolactone series (Among them AFG1, AFG2, AFGM1, AFGM2, and AFB3). The aflatoxins fluoresce strongly in ultraviolet light (ca. 365 nm); B1 and B2 produce a blue fluorescence where as G1 and G2 produce green fluorescence.

Aflatoxins B1, B2, G1, and G2 are mycotoxins that may be produced mainly by three moulds of the *Aspergillus* species:*A. flavus*, *A. parasiticus* and *A. nomius*, which contaminate plants and plant products(Bennett andKlich 2003). Aflatoxins M1 and M2, the hydroxylated metabolites of aflatoxin B1 and B2, may be found in milk or milk products obtained from livestock that has ingested contaminated feed.

Aflatoxins B1, B2, G1, G2 are the most toxic and carcinogenic naturally occurring mycotoxins. The toxic effects include acute hepatitis, immunosuppression, and hepatocellular carcinoma. In humans, the risks associated with aflatoxin consumption are well documented, and the International Agency for Research on Cancer (IARC, 1976, 1993) has designated aflatoxin as a human liver carcinogen.The aflatoxins display potency of toxicity, carcinogenicity, mutagenicity in the order of AFB1 > AFG1 > AFB2 > AFG2. Aflatoxins pose a risk to human health because of their extensive pre-harvest contamination of corn, cotton, soybean, peanuts and tree nuts, and because residues from contaminated feed may appear in milk.

3.1.2. Incidence

Surprisingly not many reports have studied the presence of aflatoxins in oats. Some of them studied their presence in very low quality grains mainly dedicated as feed and, furthermore were done many years ago when the analysis techniques were not fully developed.As one example using TLC analysis, Shotwell et al. (1969) described the presence of aflatoxin in 3 samples of oat grains. The limitation of the technique was pointed out by the same authors and they stated that maybe these results were false positives.More recently Escobar and Regueiro (2002) published a survey in which they studied the contamination of Cuban food and feedstuffs with AFB1. Samples were collected between 1990 and 1996 and they described mean levels of 10.7 μg kg^{-1} in oat grains.Buckley et al. (2007) described the absence of aflatoxin in 24 samples of oats forage and concentrated for animal feeding.In a study to evaluate mycotoxin natural occurrence in Argentinean oats used in feed (Sacchi et al., 2009) they found AFB1 in two samples out of 17 but at low concentration levels (108.0 and 105.0 μg kg^{-1}).

3.2.Ochratoxin A

3.2.1. Description and Toxicology

OchratoxinA (OTA) comprises a dihydrocoumarin moiety linked to a molecule of L-β-phenylalanine via an amide bond. The systematic chemical nomenclature for OTA is (R)-N-[(5chloro-3,4-dihydro-8-hydroxy-3-methyl-1-oxo-1H-2-benzopyran-7-yl)-carbonyl]-L-phenylalanine.

OTAis a mycotoxin produced by several fungal species in the*Penicillium* and *Aspergillus* genera, primarily *Penicilliumverrucosum*, *Aspergillus-ochraceus* and *Aspergilli* of the section *Nigri*, especially *A. carbonarius* (Medina et al., 2005). *A.ochraceus* can infect cereals, coffee, cocoa and edible nuts. *P. verrucosum* is a particularly important source of contamination in grain in cooler regionsof Northern Europe (Olsen et al. 2003).

OTA is a potent nephrotoxin, exhibits carcinogenic, teratogenic and immunotoxic properties in rats and possibly in humans (IARC 1993). The genotoxicity of OTA remains controversial (EFSA 2006). OTA is receiving increasing attention worldwide because of its wide distribution in food and feed and human exposure that most likely comes from low level of OTA contamination of a wide range of different foods (PetzingerandWeidenbach 2002). The economically most important OTA producers belong to *Aspergillus* sections Circumdati and Nigri (Samson et al. 2004; Frisvad et al. 2004).

3.2.2. Incidence

OTA occurrence in the cereal harvest from 1997 was studied Scudamore et al. (1999)in UK. In this study they analysed 21 samples of oats from different regions and they found 6 (28.6%) positive samples with a mean value for positive samples of 0.53μgkg^{-1}.

Kononenko et al. (2000) studied the presence of OTA in cereal samples from West Siberia, theUrals, and in the Kursk Region. They found two samples of contaminated oats containing 157.7 and 57.6 μg g^{-1}.

In 2001, the Joint Food and Agriculture Organization of the United Nations/World Health Organization Expert Committee on Food Additives

(JECFA) studied the safety of the presence of certain mycotoxins in food from the European countries. In their report they included results of OTA contamination of cereals. The most relevant data regarding oats is listed in Table 2.Meister (2003) developed a HPLC method for the analysis of OTA in different cereals grains. To test it analysed some oat samples and they found contamination in two samples with values of 0.7 and 1.4 μg kg^{-1}.

Table 2.Results of surveys for ochratoxinA showing concentrations and distribution of contamination in oats. Adapted from JECFA (2001)

Country/ Region	Commodity	Year/ Season	No. of samples	LOQ (μg kg^{-1})	Mean/Max (μg kg^{-1})	References
Sweden	Oat grain	1996–98	23	0.1	0.32/3.6	Thuvander et al. (2000); Larsson andMøller (1996)[b]
	Oat grain	1999	10	0.05	0.05/0.15	National Food Administration (2000); Thuvander et al. (2000); Larsson andMøller (1996)[b]
Norway	Oat groats	1990	20	0.3[a]	0.26/0.9	Langseth (1999)[b]
	Oats	1990	20	0.3[a]	0.44/5.8	
	Oats	1993	3	0.2	0.17/0.26	
	Oats	1994	3	0.2[a]	3.47/10.2	
	Oats	1995	21	0.25[a]	0.32/4.2	
	Oats	1996	14	0.3[a]	0/0	

Table 2. (Continued)

Country/ Region	Commodity	Year/ Season	No. of samples	LOQ (µgkg^{-1})	Mean/Max (µg kg^{-1})	References
	Oats	1997	14	0.01[a]	0.053/0.23	
	Oats	1998	22	0.01[a]	0.065/0.47	
	Total oats samples	1990–98	97		0.46/10.2	
Denmark	Oat kernels	1986–92	50	0.05[a]	0.5/5.6	Jorgensen et al. (1996)[c]
	Oat kernels, organic	1986–92	17	0.05[a]	0.3/4.2	
	Oat kernels, imported	1986–92	25	0.05[a]	0.5/4.6	
Finland	Oat kernels	1996	34	0.80	1.7/56.6	Solfrizzo et al. (1998)[c]
United Kingdom	Oat kernels	1997–98	21	0.1[a]	0.1/2.2	MAFF (1999)
		1996	18	0.2[a]	0.33/5.9	MAFF (1997) Sharman et al. (1992)
Germany	Oats	1995–98	30	0.01[a]	0.06/0.14	Wolff et al. (2000)

MAFF, Ministry of Agriculture, Fisheries and Food (UK).

[a] Limit of detection.

[b] Sampling not described.

[c] The number of samples was divided by a factor of 3 for calculation of the weighted mean.

4. EUROPEAN LEGISLATION

Commission Regulation (EC) No 1881/2006 of 19 December 2006 set maximum levels for certain contaminants in foodstuffs, including several mycotoxins.

All cereals and all products derived from cereals,including processed cereal products, with some exceptions should have maximum level of AFB1 of 2.0 μg kg^{-1} and total amount of all aflatoxins of 4.0 μg kg^{-1}. Also, the processed cereal-based foods and baby foods for infants and young children should have as maximum 0.10 μg kg^{-1} of AFB1.The most recent change regarding European legislation was included in February 2010. In the Commission Regulation (EU) No 165/2010 new maximum levels for aflatoxins in some food porducts were amended.

Maximum OTA levels in unprocessed cereals was set at 5.0 μg kg^{-1} and for all products derived from unprocessed cereals, including processed cereal products and cereals intended for direct human consumption maximum level was set at 3.0 μg kg^{-1}.

Regarding processed cereal-based foods and baby foods for infants andyoung children levels are more restrictive and a maximum of 0.50μg kg^{-1} is allowed.

Maximum content of DON in unprocessed durum wheat and oats are set in the same directive and maximum levels are 1750 μg kg^{-1}. In the case of Bread (including small bakery wares), pastries, biscuits, cereal snacks and breakfast cereals maximum level is 500 μg kg^{-1}. Likely,for processed cereal-based foods and baby foods for infants and young children maximum level is200 μg kg^{-1}.

Recently, the toxic effects of T-2 were taken into consideration in the Joint FAO/WHO Expert Committee on Food Additives (JECFA) where the safety of certain mycotoxins in food was evaluated (WHO/FAO, 2001). JECFA concluded that the toxic effects of T-2 and its metabolite HT-2 could not be differentiated, and that the *in vivo* toxicity of T-2 might be due partly to toxic effects of HT-2. Therefore, the provisional maximum tolerable daily intake (PMTDI) for these toxins was fixed at 60 ngkg^{-1} body weight per day, including intake of T-2 and HT-2, alone or in combination (WHO/FAO, 2001). Recently, the European Food Safety Authority (EFSA) published a report on the toxicity of these trichothecenes and they concluded that the toxicity of T-2 toxin *in vivo* is considered to include that of HT-2 toxin and the results of studies with T-2 toxin are used to approximate the effects of HT-2 toxin (Schuhmacher-Wolz et al., 2010). The European Commission (EC) has established, with Regulations No. 856/2005 and No. 1881/2006, admissible levels of several *Fusarium*toxins in cereals and cereal-based products which became effective from 1 July, 2006. Maximum admissible levels for T-2 and HT-2 toxins in unprocessed cereals and cereal products are currently under discussion (Commission Regulation (EC) No 856/2005; Commission Regulation (EC) No 1881/2006).

5. Controllingthe Emerging Menace of TheseMycotoxins

With the available data examined above, it seems that control of DON and mainly Type Atrichothecenes are of particular importance in oats. The increasing levels of T-2 and HT-2 toxins in oat grains can pose a serious health problem that should be managed.

Several data support the consideration of *F. langsethiae* as the main fungi responsible for these increasing levels of toxins in oats. As an example, molecular studies have shown that *F. langsethiae* DNA is present in ripening oats between flowering and harvest (Parikka et al., 2007; Yli-Mattila et al., 2008; Fredlund et al., 2010). Furthermore, while very poor correlations have been found between T-2+HT-2 contamination levels with species like *F. sporotrichioides* and *F. poae*, good correlations have been obtained between total *F. langsethiae* DNA and the presence of T-2 and HT-2 toxins (Yli-Mattila et al., 2008; Fredlund et al., 2010).

Environmental conditions such as relative humidity and temperature are known to have an important effect on the onset of *Fusarium*Head Blight (FHB). It has been shown that moisture conditions at anthesis are critical in *Fusarium* infection of the ears, and Lacey et al. (1999) have shown that *Fusarium* infection in the UK is exacerbated by wet periods at a critical time in early flowering in the summer, which is the optimum window for susceptibility. Equally, drought-damaged plants are more susceptible to infection.

Magan et al. (2003) studied how water activity (a_w) and temperature were affecting *F. culmorum* on growth and DON and NIV production. Production occurred in the relatively narrow a_w range 0.995-0.95 while growth persisted to 0.90 a_w. Also, the optimal conditions for DON and NIV production, 25°C at 0.995 and 0.981 a_w, respectively, were within the range optimal for growth. If we compare the above a_w levels with field samples, these $a_w$levels correspond to water contents of approximately 30% and 26%, respectively, and are within normal ranges for harvested grain in wet years. Toxin production was significantly higher at 25 °C compared to 15 °C.

As occurs with other fungal mycotoxigenic species, the conditions under which both toxins were produced were more restrictive than the conditions allowing growth of the fungus.

New data on the ecophysiology of *F. langsethiae* (Medina and Magan, 2010, 2011) suggest, as in the previous case, the high importance of water availability in determining contamination levels with these toxins. On the other hand a limited effect of the temperature has been found, and since, the toxins are produced over a wide temperature range. Thus inefficient drying may allow *F. langsethiae* to continue to colonise and increase contamination post-harvest.These studies have provided very useful data on the conditions which represent a high and low risk for contamination by these mycotoxins.

While there is more detailed information on DON, limited studies have been conducted on the impact of agronomic practices on HT-2 and T-2contamination of cereals. Edwards (2007) studied the effect of some agricultural practices in different cereals ina UK study.The results have shown that the agronomic factors are similar for HT-2 andT-2 in barley and oats, butdifferent to what is known for DON in wheat (Edwards, 2007). HT-2 and T-2 levels are higher in spring barleys in Franceand winter oats in the UK, but the reasons still unknown. These differences could be explained by genetic diversity on plants or based on prevailing weatherconditions when the crops are flowering. This stage is known to be the most pronegrowth stage for contamination and winter and spring varieties flower at different times.

Recently Parikka et al. (2007) found differences on the percentage of infected kernels depending on the cultivation practices. In this case Finnish cereals cultivated using direct drilling had higher *F. langsethiae* contamination while tilled plots were less contaminated.

Combining all these data, it is clear that different mycotoxin levels in grains would depend mainly on different inoculum/infection rates by *F. langsethiae*at field level and prevailing weather conditions in the cereal flowering stage. For reduction of T-2 and HT-2 toxins it would be recommended (Edwards, 2004, 2007; Parikka et al. 2007):

1. Grow oats after a non-cereal
2. Plough before oats
3. Grow less susceptible varieties including more spring oats
4. Use less intense cereal rotations

Once the grains are harvested, in most of the cases a storing post-harvest period start. In this moment some factors are critical. If the post-harvest conditions are poorly managed the natural microbiota carried by the grains can lead to rapid deterioration in nutritional quality ofseeds. Microbial activity can cause undesirable effects in grains including discoloration,contribute to heating and losses in dry matter through the utilization of carbohydrates as energysources, degrade lipids and proteins or alter their digestibility, produce volatile metabolitesgiving off-odours, cause loss of germination and baking and malting quality; affect use as animalfeed or as seed. And finally, as shown in this chapter, the growth of filamentous fungal spoilage organisms may lead to the accumulation ofmycotoxins(Magan et al., 2004). The spores of some fungicause respiratory disease hazards to exposed workers (Lacey and Crook, 1988).

In order to reduce all these alterations some key parameters should be controlled during the storage process. Prevention strategies nowadaysare predominantly based on using the HACCPapproach and to identify the critical control points in the pre- and post-harvest food chain. Thishas been examined for some food chains(e.g. *Fusarium* and trichothecenes) in temperate cereals(Aldred and Magan, 2004) and the HACCP approach to mycotoxin control has been reviewed by Aldred et al. (2004). From this previous knowledge we can identify as key post-harvest critical control points:

1. Regular and accurate moisture measurement determinations.
2. Efficient and prompt drying of wet cereal grain. This will be directly related to the buffer storage time and temperature prior to drying as well as the actual drying conditions (e.g. ambient, heated air drying) to target safe moisture contents. Wheat/Barley/Oats, 14-14.5%.
3. Infrastructure for quick response, including provision for segregation and appropriate transportation conditions.
4. Appropriate storage conditions at all stages in terms of moisture and temperature control, the general maintenance and effective hygiene of storage facilities for prevention of pests and water ingress.
5. Ability to efficiently identify and reject material below specified standards in terms of both fungal contamination and, at some stages, mycotoxin levels (e.g. when passing to a third party).

6. Operation of approved supplier systems. This requires the setting of specifications for acceptance/rejection.

CONCLUSION

As other small grain cereals, oats appear to be contaminated with several mycotoxins regularly. Levels found are, in most of the cases, low and with reduced frequencies but for some of the mycotoxins discussed in this chapter (T-2+HT-2 and OTA) more attention and control procedures would be necessarily implemented.

While there is more information regarding the production of OTA by *Penicilliumverrucosum* in temperate climates during storage, it still very scarce knowledge regarding whether the production of T-2 and HT-2 is pre- or post-harvest. The new ecological data shows that *F. langsethiae* would be able to produce the toxin under both conditions.

Under these conditions:

- Prevention strategies post-harvest can only be effective for mycotoxins that are formed during this component of the food chain.
- To minimize pre-harvest contamination, Good Agricultural Practice, including all new data regarding how different practices will allow the reduction of mycotoxinsare needed.
- Effective and rapid diagnostic tools which can be used to monitor and quantify these mycotoxins rapidly at early stages would be necessary.
- This control in origin is crucial since pre-harvest natural contamination can only be minimized post-harvest by application of processing techniques which will minimize the subsequent entry of the existing mycotoxin into the food and feed chain.
- There are however, as described above, key management tools and traceability procedures, which should be used to facilitate stored commodities to be effectively conserved with minimum loss in quality.
- Attention should be paid to the new global climate change scheme. Slightly elevated CO_2 concentrations and interactions with

temperature and water availability may stimulate growth of some mycotoxigenic species, especially under water stress. Also other important questions would be whether some mycotoxigenic fungi may become moreadapted to such changing stress conditions, and whether other xerophilic fungi, some producing secondary metabolite toxic to humans and animals, may become dominant in a changing environment.

REFERENCES

Aldred D. and Magan N., 2004.Prevention strategies for tricothecenes. *Toxicology Letters* 153: 165-171.

Aldred D., Olsen M. and Magan N., 2004. HACCP and Mycotoxin control in the food chain. In: *Mycotoxin in food: detection and control*, Eds. Magan, N. and Olsen, M.,Woodhead Publishing, Cambridge, U.K.

Anon. 1999.Opinion of the Scientific Committee on Food on Fusarium toxins.Part 1: Deoxynivalenol, European Commission, Brussels (http://ec. europa.eu/food/fs/sc/scf/out44_en.pdf).

Anon. 2004. Opinion of the scientific panel on the contaminants in the food chain on a request from the Commission related to deoxynivalenol (DON) as undesirable substance in animal feed. *The EFSA Journal* 73: 1-41.

Bennett J.W. andKlich M., 2003.Mycotoxins.*ClinicalMicrobiology Review 16*:497-516.

BIOMIN's Mycotoxin Survey – 3rd Quarter 2010. 2010. Accessed on-line 10-1-2011. http://www.biomin.net/en/knowledge-center/articles/articles-details/ article/biomins-mycotoxin-survey-3rd-quarter-2010/

Bottalico A.,1998.*Fusarium* diseases of cereals: Species complex and related mycotoxin profiles in Europe. Journal of Plant Pathology,80: 85-103.

Buckley T., Creighton A., Fogarty U., 2007. Analysis of Canadian and Irish forage, oats and commercially available equine concentrate feed for pathogenic fungi and mycotoxins. *Irish Veterinary Journal*, 60: 231-236.

Canady R.A., Coker R.D., Egan S.K., Krska R., Kuiper-Goodman T., Olsen M., Pestka J., Resnik S. andSchlatter J., 2001.*Safety Evaluation of Certain Mycotoxins in Food* (World Health Organization Food Additive Series 47). Geneva, Switzerland: World Health Organization; Deoxynivalenol; p. 419-555

Chulze S.N., Ramire, M.L., Farnoch, M.C., Pascal, M., Visconti A. and March G., 1996. *Fusarium* and fumonisin occurrence in Argentinian corn at different ear maturity stages.*Journal of Agriculture and Food Chemistry*, 44: 2797-2801.

Cleveland T.E., Dowd P.F., Desjardins A.E., Bhatnagar D. andCotty P.J., 2003. United States Department of Agriculture-Agricultural Research Service research on pre-harvest prevention of mycotoxins and mycotoxigenic fungi in US crops. Pest Management Science 59:629-642.

Commission Regulation (EC) No 856/2005 of 6 June 2005 amending Regulation (EC) No 466/2001 as regards *Fusarium* toxins.*Official Journal of the European Union* L 143/3-8.

Commission Regulation (EC) No 1881/2006 of 19 December 2006 setting maximum levels for certain contaminants in foodstuffs.*Official Journal of the European Union Official Journal of the European Union.*L 364/5-24.

Commission Regulation (EU) No 165/2010 of 26 February 2010 amending Regulation (EC) No 1881/2006 setting maximum levels for certain contaminants in foodstuffs as regards aflatoxins.*Official Journal of the European Union* 27.2.2010.pp 8-12

Cotty P.J., 2006.Biocompetitive exclusion of toxigenic fungi. In: *Themycotoxinfactbook: food and feed topics.*Barug D., Bhatnagar D., van Egmond H.P., van der Kamp J.W., van Osenbruggen W.A., Visconti A., eds. Wageningen:Wageningen Academic Publishers, pp.179-197.

Desjardins A.E., 2006.*Fusarium*Mycotoxins: Chemistry, Genetics and Biology. St. Paul, Minnesota, American Phytopathological Society.

D'Mello J.P.F. and Macdonald A.M.C., 1997.Mycotoxins.*Animal Feed Science Technology* 69: 155-166.

Edwards S.G., 2004. Influence of agricultural practices on *Fusarium* infection of cereals and subsequent contamination of grain by trichothecenemycotoxins. *Toxicology Letters* 153: 29-35.

Edwards S., 2007. Investigation of *Fusarium*mycotoxins in UK barley and oat production.Accessed on-line 12-1-2011.FSA CO4033 and CO4034/HGCA 2706.http://www.foodbase.org.uk/admintools /reportdocuments/48_91_CO4033_Final_report-_Investigation_of_fusarium_toxins_in_UK_.pdf

Edwards S.G., 2009. *Fusarium*mycotoxin content of UK organic and conventional oats.*Food Additives and Contaminants: Part A*, 26: 1063-1069.

EFSA 2006.European Food Safety Authority. Opinion of the Scientific Panel on contaminants in the Food Chain of the EFSA on a request from the Commission related to ochratoxinA in food. The EFSA Journal, 365: 1-56. Available from: http://www.efsa.europa.eu/etc/medialib/efsa/-science/

contam/contam_opinions/1521.Par.0001.File.dat/contam_op_ej365_ochratoxin_a_food_en1.pdf

Eriksen G.S. and Alexander J. (eds.), 1998.*Fusarium* toxins in cereals - a risk assessment.Nordic Council of Ministers; *Tema Nord*, 502, Copenhagen pp. 7-44.

Escobar, A.andRegueiro, O.S., 2002. Determination of Aflatoxin B1 in food and feedstuffs in Cuba (1990 through 1996) using an immunoenzymatic reagent kit (Aflacen). *Journal of Food Protection*, 65: 219-221.

Flannigan, B., 1991. Mycotoxins. In: D'Mello, J.P.F., Duffus, C.M., Duffus, J.H. (Eds.), *Toxic Substances in Crop Plants. The Royal Society of Chemistry,* Cambridge, pp. 226-257.

Food Standard Agency, UK. Mycotoxins in oats survey. 2004. Accessed on-line 10-1-2011. http://www.food.gov.uk/news/ newsarchive/ 2004/ feb/ 174899

Fredlund E., Gidlund A., Pettersson H., Olsen M. andBörjesson T., 2010.Real-time PCR detection of *Fusarium* species in Swedish oats and correlation to T-2 and HT-2 toxin content.*World Mycotoxin Journal*, 3: 77-88.

Frisvad J.C., Frank J.M., Houbraken J.A.M.P., Kuijpers A.F.A. and Samson R.A., 2004. New ochratoxinA producing species of *Aspergillus* section *Circumdati*. *Studies in Mycology50*:23-43.

Gottschalk C., Barthel J., Engelhardt G., Bauer J. and Meyer K., 2007. Occurrence of type Atrichothecenes in conventionally and organically produced oats and oat products. *Molecular Nutritionand Food Research*, 51: 1547-1553.

Hietaniemi V. andKumpulainen J., 1991. Contents of *Fusarium* toxins in Finnish and imported grains and feeds.*Food Additives and Contaminants*, 8: 171-182.

Hudec, K. andRohácik, T., 2009.The occurrence and predominance of *Fusarium* species on barley kernels in Slovakia.*Cereal Research Communications,* 37: 101-109.

IARC Monographs on the Evaluation of Carcinogenic Risks to Humans. 1976. Some Naturally Occurring Substances. IARC Monographs on the Evaluation of Carcinogenic Risk of Chemicals to Humans, vol. 10. Lyon, *France:International Agency for Research on Cancer*. pp.353

IARC Monographs on the Evaluation of Carcinogenic Risks to Humans. 1993. Some Naturally Occurring Substances: Food Items and Constituents, Heterocyclic Aromatic Amines, and Mycotoxins. *IARC Monographs onthe Evaluation of Carcinogenic Risk of Chemicals to Humans*, vol. 56.IARC press, Lyon, France.pp.571

IARC Monographs on the Evaluation of Carcinogenic Risks to Humans (1993).Ochratoxin A. In: *Some Naturally Occurring Substances: Food Items and Constituents, Heterocyclic Aromatic Amines and Mycotoxins*. IARC Press, Lyon, France 56: 489 -521.

Joint Food and Agriculture Organization of the United Nations/World Health Organization Expert Committee on Food Additives (JECFA). 2001. *Safety evaluation of certain mycotoxins in food.* WHO Food Additives series 47; FAO food and nutrition paper 74.[Online.] http://www.inchem.org/documents/jecfa/jecmono/v47je01.htm.

Johnsen H., Odden E., Johnsen B. A., Beyum A. and Amundsen, E., 1988. Cytotoxicity and effects of T2-toxin on plasma proteins involved in coagulation, fibrinolysis and kallikrein-kinin system. *Archives of Toxicology,* 61: 237-240.R.

Jorgensen K., Rasmussen G. and Thorup, I., 1996.Ochratoxin A in Danish cereals 1986-1992 and daily intake by the Danish population.*Food Additives and Contaminants*, 13: 95-104.

Kononenko G.P., Burkin A.A., Zotova E.V., and Soboleva N.A., 2000.Ochratoxin A: Contamination of Grain. *Applied Biochemistry andMicrobiology,* 36: 177-180.

Kosiak B., TorpM. and Thrane U.L.F., 1997.The occurrence of *Fusarium* spp. in Norwegian grain-A survey. *Cereal Research Communications*, 25: 595-596.

Kozakiewicz Z., 1989. *Aspergillus* species on stored products.*Mycological Papers,*161: 1-188.

Lacey J. and Crook B. 1988. Fungal and actinomycete spores as pollutants of the workplace and occupational allergens.*Annals of OccupationalHygiene,* 32: 515-533.

Lacey J., Bateman G.L. and Mirocha C.J., 1999.Effects of infection time and moisture on the development of ear blight and deoxynivalenol production by *Fusarium* spp. in wheat.*Annals of Applied Biology*,134: 277-283.

Langseth W. andElen O., 1996.Differences between barley, oats and wheat in the occurrence of deoxynivalenol and other trichothecenes in Norwegian grain.*Journal of Phytopathology*, 144: 113-118.

Langseth W. andRundberget T., 1999.The occurrence of HT-2 toxin and other trichothecenes in Norwegian cereals.*Mycopathologia*, 147: 157-165.

Larsson K. and Moeller T., 1996. Liquid chromatographic determination of ochratoxinA in barley, wheat bran, and rye by the AOAC/IUPAC/NMKL method: NMKL collaborative study. *Journal AOAC International*,79: 1102-1105.

Magan N., Hope R., Cairns V. and Aldred D., 2003. Post-harvest fungal ecology: impact of fungal growth and mycotoxin accumulation in stored grain.*European Journal of Plant Pathology*, 109: 723-730.

Magan N., Sanchis V. and Aldred D., 2004.Role of spoilage fungi in seed deterioration. Chapter 28, In: *Fungal Biotechnology in Agricultural, Food and Environmetal Applications*, Edt. D.K.AuroraMarcell Dekker. Pp. 311-323.

Matthaus K., Danicke S., Vahjen W., Simon O., Wang J., Valenta H., Meyer K., Strumpf A., Ziesenib H. andFlachowsky G., 2004. Progression of mycotoxin and nutrient concentrations in wheat after inoculation with *Fusariumculmorum*.*Archives of Animal Nutrition*58: 19-35.

Mankevičienė A., Butkutė B., Dabkevičius Z. andSupronienė S., 2007.*Fusarium*mycotoxins in Lithuanian cereals from the 2004-2005 harvests.*Annals of Agricultural and Environmental Medicine*, 14: 103-107.

Matsumoto H., Ito T. andUeno Y., 1978.Toxicological approaches to the metabolites of fusaria. XII. Fate and distribution of T-2 toxin in mice. *Japanese Journal of Experimental Medicine,* 48: 393-399.

Medina A., Mateo R., Lopez-Ocana L., Valle-Algarra F. M. and Jimenez M., 2005. Study of Spanish grape mycobiota and ochratoxinA production by isolates of *Aspergillustubingensis* and other members of *Aspergillus* section Nigri.*Applied and Environmental Microbiology*, 71: 4696-4702.

Medina A. and Magan N., 2010.Effect of water activity and temperature on growth of *Fusariumlangsethiae* strains from northern Europe. *InternationalJournal of Food Microbiology* 142: 365-9.

Medina A. and Magan N., 2011.Water availability and temperature affects production of T-2 and HT-2 by *Fusariumlangsethiae* strains from north European countries. *Food Microbiology*28: 392-398.

Meister U., 2003. Detection of Citrinin in Ochratoxin A-containing Products by a New HPLC *Method Mycotoxin Research*, 19: 27-30.

Ministry of Agriculture, Fisheries and Food, 1997.*Survey of Aflatoxins and Ochratoxin A in cereals and retail Products* (Food Surveillance Information Sheet No. 130), London: Joint Food Safety and Standard Group.

Ministry of Agriculture, Fisheries and Food, 1999.*Survey of Ochratoxin A in grain traded by central deposits 1997-1998* (Food Surveillance Information Sheet No. 171), London: Joint Food Safety and Standard Group.

Muller H.M. andSchwadorf K., 1993.A survey of the natural occurrence of Fusarium toxins in wheat grown in a south-western area of *Germany. Mycopathologia*,121: 115-121.

Parikka, P., Hietaniemi, V., Rämö, S. andJalli, H., 2007. The effect of cultivation practices on *Fusariumlangsethiae* infection of oats and barley. *In:*Vogelgsang, S., Jalli, M., Kovács, G. and Vida, G. (Eds). *Proceedings of the COST SUSVAR.Fusarium subgroup meeting: Fusarium diseases in cereals-potential impact from sustainable cropping systems.* 1-2 June 2007. Velence, Hungary

Pettersson H., Borjesson T., Persson L., Lerenius C., Berg G. andGustafsson G., 2008.T-2 and HT-2 toxins in oats grown in northern Europe.*Cereal Research Communications*, 36:591-592.

Petzinger E., Weidenbach A., 2002. Mycotoxins in the food chain: the role of ochratoxins.*Livestock Production Sciences*, 76:245-250.

Roscoe, V.,Lombaert, G.A., Huzel, V., Neumann, G., Melietio, J., Kitchen, D., Kotello, S.,Krakalovich, T., Trelka, R. and Scott, P.M., 2008.Mycotoxins in breakfast cereals from the Canadian retail market: A 3-year survey'. *Food Additives and Contaminants: Part A*, 25: 347-355.

Sacchi C., González H.H.L., Broggi L.E., Pacin A., Resnik S.L., Cano G. andTaglieri D., 2009. Fungal contamination and mycotoxin natural occurrence in oats for race horses feeding in Argentina. *Animal FeedScience and Technology*, 152: 330-335.

Samson R.A., Houbraken J.A.M.P., Kuijpers A.F.A., Frank J.M., Frisvad J.C., 2004. New ochratoxin A or sclerotium producing species in *Aspergillus* section *Nigri*. *Studies in Mycology50*: 45-61.

Schothorst, R.C. and van Egmond, H.P., 2004. Report from SCOOP task 3.2.10 "collection of occurrence data of *Fusarium* toxins in food and assessment of dietary intake by the population of EU member states": Subtask: trichothecenes. *Toxicology Letters*, 153: 133-143

Schuhmacher-Wolz, U., Heine, K. and Schneider, K., 2010. Report on toxicity data on trichothecenemycotoxins HT-2 and T-2 toxins. CT/EFSA/ CONTAM/2010/03. Available online: http://www.efsa.europa.eu/en/scdocs/scdoc/65e.htm?WT.mc_id=EFSAHL01andemt=1

Scudamore K.A., Patel S. and Breeze V., 1999. Surveillance of stored grain from the 1997 harvest in the United Kingdom for ochratoxin A. *Food Additives and Contaminants: Part A*, 16: 281-290

Sharman M., MacDonald S. and Gilbert J., 1992.Automated liquid chromatographic determination of ochratoxin A in cereals and animal products using immunoaffinity column clean-up.*Journalof Chromatography,*603: 285-289.

Shotwell O.L., Hesseltine C.W., Burmeister H.R., Kwolek W.F., Shannon G.M., and Hall H.H., 1969. Survey of cereal grains and soybeans for the presence of Aflatoxin. I. Wheat, grain sorghum, and oats. *Cereal Chemistry,* 46:446-453.

Solfrizzo M., Avantaggiato G. and Visconti A., 1998.Use of various clean-up procedures for the analysis of ochratoxin A in cereals.*Journal of ChromatographyA,*815: 67-73.

Tanaka T., Yamamoto S., Hasegawa A., Aoki N., Besling J.R., Sugiura Y. and Ueno Y., 1990. A survey of the natural occurrence of *Fusarium* mycotoxins, deoxynivalenol, nivalenol and zearalenone, in cereals harvested in The Netherlands.*Mycopathologia*, 110: 19-22.

Tanaka T., Hasegawa A., Yamamoto S., Lee U.-S., Sugiura Y., and Ueno Y., 1988. Worldwide contamination of cereals by the *Fusarium* mycotoxins Nivalenol, Deoxynivalenol, and Zearalenone. 1. Survey of 19 Countries. *Journal of Agricultural and Food Chemistry*, 36: 979-983.

Torp M. and Adler A., 2004. The European *Sporotrichiella* Project: A polyphasic approach to the biology of new *Fusarium* species. *International Journal of Food Microbiology* 95, 257-266.

Torp M. and Langseth W., 1999.Production of T-2 toxin by a *Fusarium* resembling *Fusariumpoae.Mycopathologia*, 147: 89-96.

Torp M. and Nirenberg H.I., 2004.*Fusariumlangsethiae* sp. nov.on cereals in Europe.*International Journal of Food Microbiology* 95: 247-256.

Thrane U., Adler A., Clasen P.E., Galvano F. andLangseth W., 2004.Diversity in metabolite production by *Fusariumlangsethiae*, *Fusariumpoae*, and *Fusariumsporotrichioides.International Journal of Food Microbiology*, 95: 257-266.

Thuvander, A., Möller, T., Enghart-Barbieri, H., Jansson, A., Salomonsson, A.-C.and Olsen, M., 2001.Dietary intake of some important mycotoxins by the Swedish population.*Food Additives and Contaminants* 18: 696-706.

Varga J., Juhász A., Kevei F. andKozakiewicz Z., 2004.Molecular diversity of agriculturally important *Aspergillus* species.*European Journal of PlantPathology*, 110:627 -640

Viquez O.M., Castell-Perez M.E. and Shelby R.A., 1996.Occurrence of fumonisin B1 in maize grown in Costa Rica.*Journal of Agricultural andFood Chemistry*, 44: 2789-2791.

Visconti A., 2001. Problem associated with *Fusarium*mycotoxins in cereals. *Bulletin of the Institute for Comprehensive Agricultural Sciences*, Kinki University 9: 39-55.

WHO/FAO, 2001.Safety evaluation of certain mycotoxins in food.Prepared by the fifty-sixth meeting of the Joint FAO/WHO Expert Committee on Food Additves (JECFA).WHO food additives series 47. FAO food and nutrition paper 74. International Programme on Chemical Safety, World Health Organization, Geneva

Wolff J., Bresch H., Cholmakow-Bodechtel C., Engel G., Erhardt S., Gareis M., Majerus P., Rosner H. andScheuer R.,2000. *Burden of ochratoxinAin food and in the consumer*.Final report. Institute for the Biochemistry of Cereals and Potatoes. Federal Institute for Research on Cereals, Potatoes and Fat (in German).

Yli-Mattila T., Paavanen-Huhtala S., Parikka P., Hietaniemi V., Jestoi M., Gagkaeva T., Sarlin T., Haikara A., Laaksonen S. and Rizzo A., 2008. Real-time PCR detection and quantification of *Fusariumpoae*, *F.graminearum*, *F. sporotrichioides* and *F. langsethiae* as compared to mycotoxin production in grains in Finland and Russia.*Archives of Phytopathology and Plant Protection*, 41: 243-260.

In: Oats: Cultivation, Uses and Health Effects ISBN 978-1-61324-277-3
Editor: D. L. Murphy, pp. 125-146

Chapter 4

TAILORING OAT FOR FUTURE FOODS

Asif Ahmad[a] and Zaheer Ahmed[b*]
[a]Department of Food Technology, Pir Mehr Ali Shah Arid Agriculture University Rawalpindi, Pakistan
[b]Department of Home and Health Sciences, Allama Iqbal Open University, Islamabad, Pakistan

ABSTRACT

Oats (*Avena sativa* L.) is one of the important crops of arid and semi arid zone and is rich in polysaccharides, vitamins proteins, phenolic compounds and antioxidants compounds. Especially its component, β-glucan has got considerable importance because of its numerous industrial, nutritional and health benefits. It has a great prospective to find its application in nutraceutical food industry in this century that is driven by convenience, health, taste and sustainable production. But unfortunately, at present most of this crop is being used only as feed for animals and its actual potential is yet to be explored for industrial purposes. Different food agencies and research organizations are in the process of recognizing health benefits of oats, it is therefore timely to contemplate potential of oat crop for developing nutraceutical industry in this millennium.

Important functional ingredients in oats are dietary fiber, vitamins, proteins, phenolic compounds, antioxidants and a special nutraceutical ingredient known as β-glucan (β-1→3, 1→4 glucan). The food industry

[*] Corresponding Author: (Zaheer Ahmed), Tel:++92519057265; Email: zaheer_863@yahoo.com

can take advantage of the unique properties of oat β-glucan, to use these for production of new food products. But extraction and purification of β-glucan involves a complex process and require special attention to capitalize on its yield and functional properties.

Previous research on β-glucan has demonstrated its multiple human health benefits. Such as its tendency to reduce onset of colorectal cancer, increased stool bulk, mitigate constipation, reduction in glycemic index, flattening of the postprandial blood glucose levels and insulin rises. Keeping in view of its health benefits, a range of functional foods containing β-glucan are now being commercially introduced to the market. Thus, the industrial demand for this natural cereal based compound is fast growing and has a great potential in future foods. This chapter contains details about significance of nutraceutical components from oat, β-glucan extraction techniques, β-glucan yield and recovery, chemical analysis of β-glucan, β-glucan characterization, functional properties of β-glucan, FTIR analysis of β-glucan and how β-glucan can influence glucose and lipoprotein profile to provide health benefits.

1. INTRODUCTION

Oats (*Avena sativa L.*) is an ancient traditional cereal crop grown in all parts of the world and have huge consumption as feed and food in developing countries. Although this cereal crop is in use in daily diet since primeval history yet its beneficial ingredients were not well characterized till this time. Much of the recent researches in past two decade ascertain the health benefits and nutraceutical importance of its ingredients. The major nutraceutical characteristics in this crop arise due to presence of dietary fiber. The most important dietary fiber associated with oats is β-glucan. This Dietary Fiber (DF) of oat offer a resistance against digestion in small intestine of mono-gastric animals including humans but part of this can be partly fermented in large intestine that largely dependent on solubility of DF. Thus the dietary fiber may be classified in two major classes according to the solubility identified as Insoluble (IDF) Dietary Fiber and Soluble Dietary Fiber (SDF). Insoluble DF mainly contain non-starch polysaccharides with small amounts of lignin and can retain high volume of water resultantly it provide volume to feaces, affecting bowel transit thus reduce the risk of colorectal cancer (Bingham, 1990; Hill, 1997). Soluble dietary fiber of oat is mainly composed of ß-glucan, arabinoxylan and some minor minerals. In small intestine this soluble fraction of dietary fiber tends to increase the viscosity of food contents thus delays gastric emptying with reduction in intestinal transit time (Wisker *et al*., 2000) as a result of these pheneomenon glucose and sterol absorption advantageously slow down in the intestinal tract (Wood *et al*., 1990; Kahlon and Chow, 1997).

Previous research on SDF and IDF of oat has established its manifold benefits on human health, such as its tendency to reduce glycemic index (Cavallero *et al*., 2002; Jenkins, *et al*., 2002), increased stool bulk, mitigation of constipation (Odes *et al*., 1993; Valle-Jones, 1985), delays in onset of colorectal cancer (Dongowski *et al*., 2002), and protection against sharp increase in insulin with balancing of postprandial blood glucose levels (Hallfrisch *et al*., 2003; Li *et al*., 2003). Keeping in view of its health benefits, a range of functional foods containing ß-d-glucan are now being commercially introduced in the markets world wide ([Burkus and Temelli, 2000; Temelli, 2001) with the objective to lessen the chance of cardiovascular diseases, lowering of blood cholesterol (Newman *et al*., 1992; Kahlon *et al*., 1993; Behall *et al*., 1997; Keogh *et al*., 2003), reduction in problem of obesity (Bourdon *et al*., 1999), hypercholesterolemia (Maki *et al*., 2003; Yang *et al*., 2003), better control on diabetes (Wood, 1993; Newman *et al*., 1992; Brennan and Tudorica, 2003; Pick *et al*., 1996), cancer (Sier *et al*., 2004), hypertension (Anderson, 1983; Anderson, 1990) and support in growth of beneficial intestinal micro flora (Crittenden *et al*., 2002; Tungland, 2003).

Last two decay is the magnificent period for research and acceptance of ß-glucan on industrial level. Its popularity increased as major functional, bioactive ingredients in some of the food processes especially in cereal-based foods ([Lazaridou and Biliaderis, 2007). Based on unique properties of oat ß-glucan, the food industry exploited the opportunity to develop new food products by adding this valuable ingredient into various food products including: breakfast cereals, beverages, bread and infant foods (Flander *et al*., 2007, Yao *et al*., 2007). It can also improve the nutritive quality of food and may have positive health benefits (FDA, 1997). Thus, the industrial demand for this natural cereal-based compound is fast growing and has a great potential in future foods.

Numerous factors may affect the recovery of ß-glucan from their sources, this make the extraction process a complex task. Among these factors, pH and temperature exert greater influence on chemical composition, yield and recovery of ß-glucan. By keeping these parameters variable, processor may modify the nature and composition of extracted dietary fiber and ß-glucan that may ultimately direct its specific use in food products. This will also influence the health related issues of dietary fiber especially ß-glucan. This underline the need for better extraction and purification technologies of ß-glucan and require special attention to capitalize on its yield and functional properties. Previously some extraction methodologies of ß-glucan from barley and oat were developed by several scientists (Wood *et al.*, 1977; Wood *et al.*, 1978; Bhatty, 1993, Bhatty 1995; Dawkins, 1993; Temelli, 1997) and in recent years these extraction technologies were modified by Irakli *et al*; (2004) and Ahmad *et al.*, (2009) by introduction of various enzymes in extraction processes. The outcome of this research indicated that temperature and pH are the significant factors that influence recovery of β-glucan fiber as well as functional properties of extracted ß-glucan.

Today there is no doubt that ß-glucan can offer many nutritional and rheological advantages to the food products, and food industry always shows a keen interest in physiochemical and functional properties of novel nutraceutical compounds, as this will help in the choice of specific type of compound with particular characteristics for a specific food product.

2. RESULTS

2.1. Chemical Analysis of Raw Material

Samples of oat grains (cultivar Avon) was procured from Fodder research institute Sargodha, Samples was cleaned and milled using high speed pin mill. Ground and homogenized samples were passed through a screen having openings of 0.5 mm size. The chemical analysis of the ground oat samples were performed by adopting reference as described in AACC (2000). For analysis of ß-glucan, SDF, IDF and TDF analytical kits from Megazyme International Ltd, Wicklow, Ireland was used.

Chemical analysis of oat flour (Figure 1) indicated that it is a good source of dietary fiber (DF) and presence of higher amounts of DF in this cultivar showed its potential to be used as functional ingredient in many foods. The TDF (11.09%), IDF (8.04%) and SDF (3.05%) content of tested oat cultivar in present study showed its prospective for extraction of dietary fiber from this cultivar but presence of ash, protein, crude fat in composition of oat flour suggested for a comprehensive extraction procedure for maximum extraction of dietary fiber. The major portion of this dietary fiber is in the form of β-glucan. The chemical composition of oat flour also indicated that starch (57.4 %), and protein (10.81 %) is the major impurities in oat flour that must be removed to enhance the recovery of dietary fiber in the shape of ß-glucan.

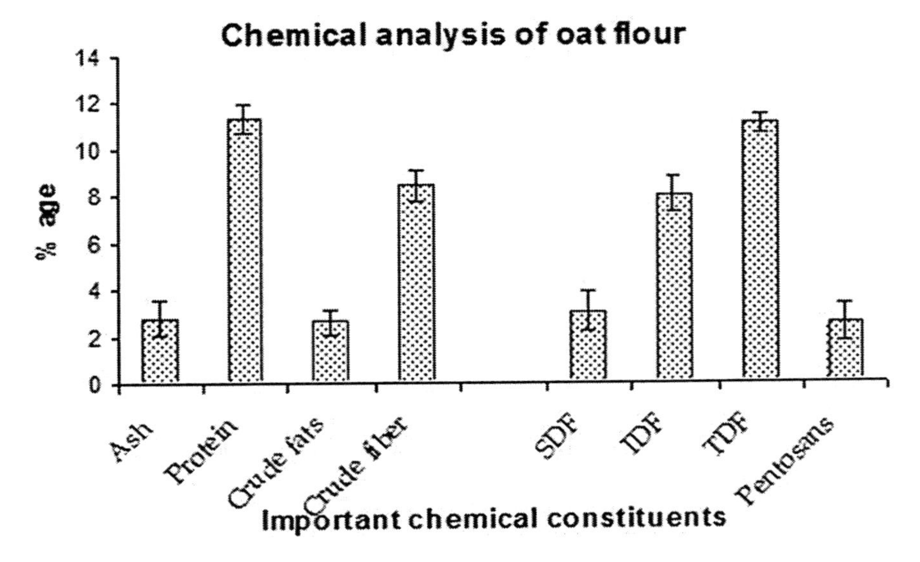

Figure 1. Chemical analysis of oat Flour.

2.2. β-Glucan Extraction

During extraction of ß-glucan from oat, flour protein and starch appeared as major impurities in the flour. These impurities have a great bearing on selection of appropriate extraction method.

To enhance purity of ß-glucan from oat flour, it is vital to remove these impurities at its maximum. For this purpose the samples of whole oat flour were refluxed through 80% ethanol and were heated with 1M NaOH to inactivate the native enzymes. Further impurities were removed by treating the supernatant in three different ways.

In first method acidic condition was maintained by using citric acid, In 2nd method impurities was removed in alkaline conditions by the use of Na_2CO_3 and in 3rd method enzymes were used for the extraction and purification of β-glucan. A schematic outline of the extraction procedure is presented in Figure 2.

All of these methods yield β-glucan in the form of gum pellets. Extracted β-glucan gum pellets were characterized for various physicochemical and functional properties as discussed in later part of this chapter.

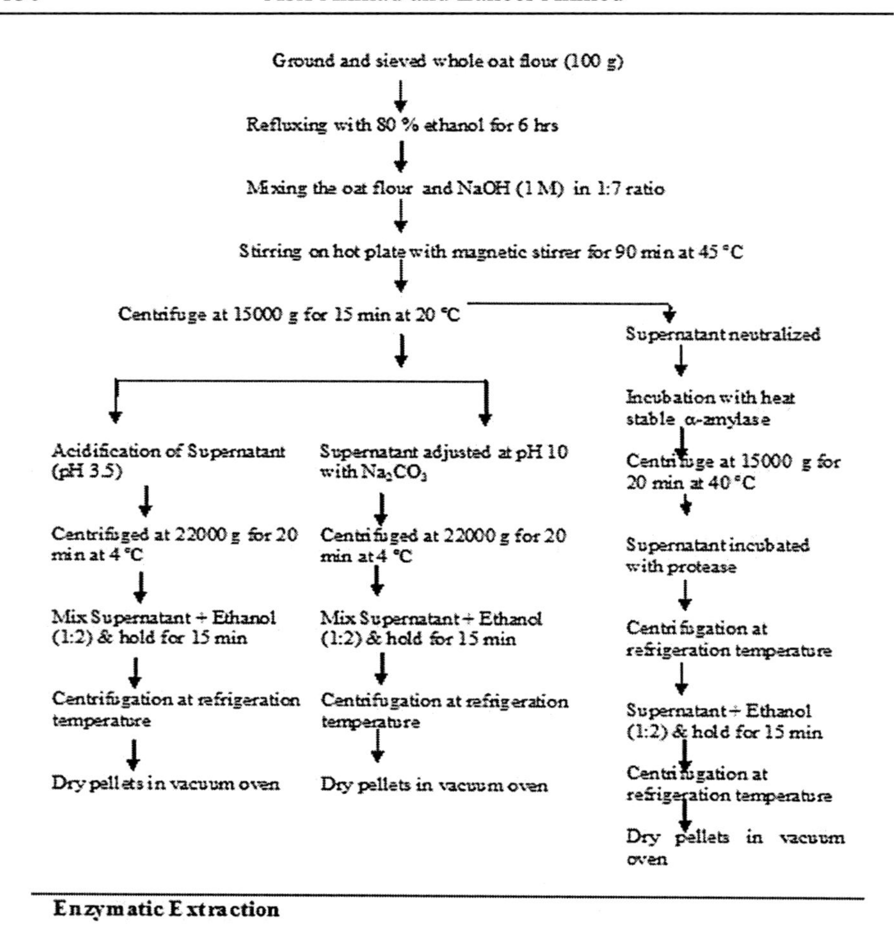

Figure 2. Schematic protocol for extraction and purification of β-glucan from oat.

2.3. β-Glucan Yield and Recovery

Yield of gum is a measure of the gum pellets that was obtained from 100 g of oat flour. Extraction procedure affected the gum yield significantly ($P<0.05$) that ranged between 3.74 and 5.14 %. Highest yield of β-glucan (5.14%) was achieved through enzymatic extraction process (Figure 3). Minor amount of fat, protein, starch, pentosans and minerals (ash) matter was also extracted along with β-glucan. Concomitant extractions of these constituents actually influence the physiochemical, nutritional and functional properties of extracted β-glucan gum pellet.

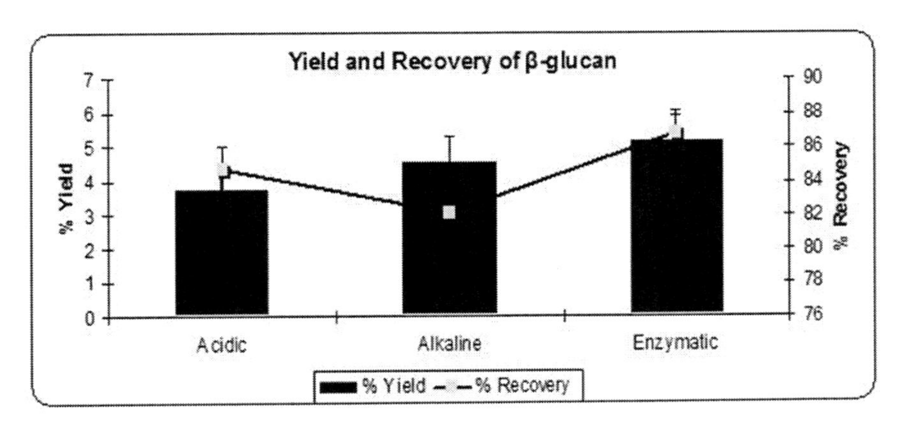

Figure 3. Yield and recovery of β –glucan.

At the same time, extraction of these constituents results in lowering of the recovery of β-glucan in the gum pellets. Several researchers also reported extraction of small amounts of protein, starch and other chemical constituents along with (Wood 2002; Burkus and Temelli, 2005) that reduce the recovery of β-glucan during extraction process.

Consequently, in order to determine the efficiency of various extraction methods recovery of β-glucan was calculated. β-glucan recovery represent the % ratio of weight of β-glucan in extracted gum product to the weight of β-glucan in 100 g flour and it is the indicator of purity of β-glucan extracted through different extraction procedures. The recovery of β-glucan ranged between 82.1 and 86.8 %. Highest recovery was obtained in enzymatic extraction method (86.8%) and lowest was observed in alkaline extraction method. Higher recovery of β-glucan in enzymatic extraction process was due to more removal of starch and protein by their respective enzymes. A prior treatment with ethanol was also decisive factor for increased recovery in enzymatic extraction procedure. This was reinforced with use of enzyme that reduces the intermolecular association of β-glucan with other components of oat that resulted in the highest extractability of β-glucan in such treatments.

2.4. β-Glucan Characterization

2.4.1. Chemical Analysis of Extracted β-Glucan Gum

Chemical analysis revealed protein (Figure 4) as major impurity that was extracted along with β-glucan in all extraction procedures. Statistically all extraction procedures varied significantly (P<0.05) with each others for all chemical constituents.

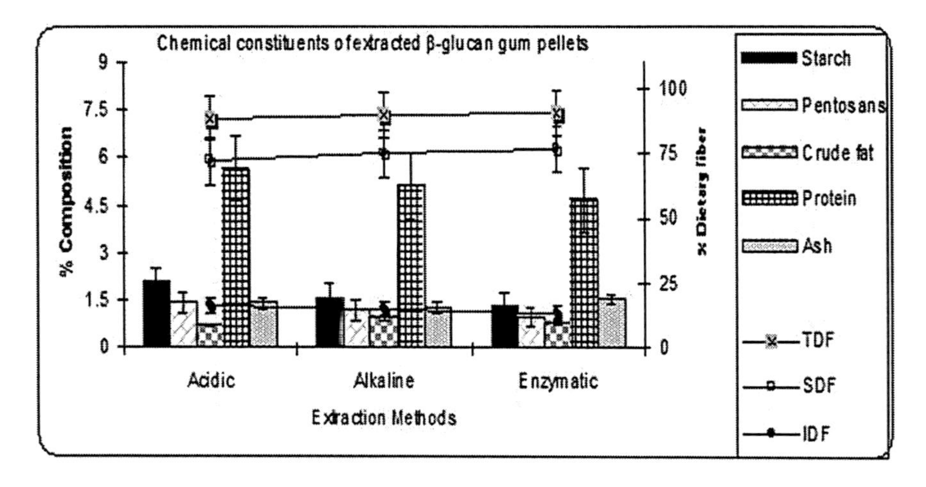

Figure 4. Chemical analysis of β-glucan gum pellets.

Comparing all extraction methods for protein impurities, enzymatic extraction process appeared more efficient with least amount of protein in β-glucan gum pellets. This extraction technique also removed efficiently greater quantities of starch, fat and pentosans. The low levels of these impurities was due to action of heat stable alpha amylase and protease enzyme that efficiently remove these impurities from extracted β-glucan gum. Acidic extraction method yielded higher amounts of starch and pentosans as impurities in extracted β-glucan gum pellets.

In alkaline extraction procedure crude fat impurity was in highest amount as compared to other extraction methods of β-glucan. Pentosans has an application in many food preparation processes and its higher amount in β-glucan can be used advantageously.

These may influence the bread properties by slowing down the rate of staling through interaction with other components. Change in pH of extraction medium is the major factor that causes variability in content of pentosans in extracted β-glucan gum pellets. Overall, enzymatic extraction process appeared to more efficient in removing impurities during extraction of β-glucan gum. Dietary fiber (TDF, SDF, IDF) varied significantly (P<0.05) in all tested extraction methods. Higher amounts of SDF and TDF was extracted in gum pellets samples that was extracted by enzymatic extraction process.

Extracted gum pellets contain relatively less amount of IDF that indicated its application as thickener, water binding substance and this lead to its use in many food industries. The importance of dietary fiber is manifested in lowering of total and LDL cholesterol. It was estimated that 1 gram of SDF from oats could lower 0.037 mmol/L of total cholesterol (Brown *et al.*, 1999). Industrially dietary fiber of oat can be used for manufacturing of high fiber bread and biscuits.

2.4.2. Minerals Profile of Extracted β-Glucan

It's a universal accepted fact that dietary fiber make complex with divalent and monovalent metals. This chelating process varies according to nature and composition of dietary fiber and offers a buffering action in the solutions. These minerals also tends to modify physiochemical and functional properties of dietary fiber. Minerals contents in present study were determined by adopting the reference protocol as outlined in AOAC (1990) , For this purpose Atomic Absorption spectrophotometer (Varian, AA240, Victoria, Australia) was used. The results of this research indicated the dominant minerals were phosphorus and potassium that was extracted along with β-glucan (Table1).

Medium amounts of magnesium and calcium was also extracted in gum pellets. Presence of these minerals is very important in buffering action as well as providing binding characteristics to gum pellets. Minor amounts of Iron, manganese and copper was also observed.

The small variation in tested extraction methods may be due to change in pH of the extraction medium. Extraction of these minerals not only improve the nutritional importance of these gum pellets but also resulted in improvement of functional properties of these gum pellets that leads its utilization in many industrial processes.

Table 1. Minerals contents of β-Glucan in gum pellets (mg /Kg)

	P	K	Mg	Ca	Na	Zn	Fe	Mn	Cu
M1	3245	3129	873	564	198	95.3	45	9.4	5.1
M2	2987	3145	865	459	172	56	51	6.8	5.3
M3	3086	3045	812	369	124	48.9	65	8.6	5.8

All values are means of three replicates.

M1=Acidic Extraction Method; M2=Alkaline Extraction Method; M3=Enzymatic Extraction Method.

2.5. Functional Properties

2.5.1. Water Binding Capacity

For determination of Water binding capacity 200 mg samples of β-glucan was taken into centrifuge tubes and initially shake on a digital shaker at 25 ◦C for 3 h after which it was centrifuged at 14,000×g for 30 min at 25 ◦C. The supernatant (unbound water) was discarded, and the amount of water held in the hydrated sample was determined by heating the pre-weighed pellet in a hot air oven for 2h at 120◦C. The water binding capacity of each sample was expressed as the weight of water held by 1.0 g of β-glucan sample.

The results of the water binding capacity (WBC) is presented in Table 2. The WBC of the β -glucan ranged between 3.14 and 4.52 gg^{-1} in all samples extracted through various extraction procedures. All of the extraction processes exhibited a significant variation with respect to water binding capacity of β-glucan gum pellets. β-glucan extracted from acidic extraction procedure showed highest water biding capacity. This higher WBC was attributed to β-glucan and pentosan content of extracted gum. Higher water binding capacity in extracted β-glucan makes it ideal to be used as stabilizer and as anti synersis ingredient in jams, jellies, marmalades and spreads.

Table 2. Functional properties of β-Glucan

	Water Binding Capacity g/g	Whippibility (%)	Foam stability (%)	Viscosity (cp)
M1	4.52 a	172	60.2	35.6
M2	3.14 c	158	58.5	40.84
M3	3.95 b	182	61.8	56.16
CV (%)	1.68	3.54	2.85	2.65
LSD	0.96	6.38	7.15	1.24

M1=Acidic Extraction Method; M2=Alkaline Extraction Method; M3=Enzymatic Extraction Method.

2.5.2. Whippibilty and Foaming Stability

Whippibility is important in many food-processing operations such as in cake formation and was determined as a function of increase in volume. Small amount of extracted β-glucan gum (2.5 g) was dissolved in 100 ml distilled water and shaken vigorously for 2min using a hand held food mixer at high speed in a stainless steel bowl with straight sides and volumes were recorded before and after whipping. The percentage volume increase (which serves as index of foaming capacity or whippability) was calculated as percent change in volume of solution before and after mixing. To determine the foaming stability (FS), developed foams were slowly transferred to a 1000 ml graduated cylinder and the volume of foam that remained after staying at 25±2 for 2 h was expressed as a percentage of the initial foam volume.

Whippibility of β-glucan gum pellets was appeared to be affected by extraction technique (Table 2). A pH based dependency was observed in whippibility of gum pellets that was significant in all extraction procedures. Alkaline extraction procedure resulted in reduced capacity in whippibility of β-glucan gum pellets. The tendency to form better and stable foams of β-glucan gum pellets was due to establishment of stable network with the ability to hold smaller solute particles. A nin significant influence of ext6raction techniques was observed on foaming stability of gum pellets. Despite of the fact, that with passage of time there was a small decrement in the developed foam yet more than 50% of the foam stability was observed even after two hours of the foam devlopment in all samples of β-glucan gum pellets irrespective of the method it was extracted.

2.5.3. Viscosity

For viscosity measurements, dispersion of β-glucan gum pellets was made by dissolving β-glucan gum pellets (1% w/v, as is basis) in deionized water and heating this mixture at100 ◦C for 10 min followed by stirring on magnetic stirrer at 30 °C for 2 h and adjusting pH at 7.0. Present study showed a significant variation in viscosity of β-glucan gum pellets that ranged between 35.6 and 56.16 cp. Samples of β-glucan gum pellets extracted through enzymatic extraction procedure exhibited higher viscosities. Whereas acidic and alkaline extraction procedure tends to produce gum pellets that was characterized with lower viscosity. The reduced viscosity of samples extracted through alkaline extraction process was attributed to disturbances in (1→3)-b-d bond at higher pH. These 1→3 linkages remain intact when enzymatic extraction process was used. Therfore, this process is proved to provide less harsh effects to chemical nature of β-glucan that resulted in increased viscosity of gum pellet.

Increased viscosity was also reinforced by existence of some non-β-glucan ingredients such as pentosans and some viscosity modifying metal ions in the extracted β-glucan. Characteristically high viscosity of β-glucans gum pellets can be used advantageously as stabilizer and thickening agents in various food products such as beverages, ice creams, dips, jellies, sauces, cake mixes and salad dressings.

2.6. Color of β-Glucan Gum Pellets

The LAB system of color measurement was employed to determine the color of β-glucan gum pellets samples. In this system, the values for L* denotes the lightness of the sample, these values range between L* = 100 as white and L* = 0 as black; It is athree dimensional system having coordinates for colors i.e a* = red (+) or green (−) direction; b* = yellow (+) or blue (−) direction of a product. The units within the *L*, a*,b** system give equal perception of color difference to a human observer.

In this study L* value of β- glucan gum pellet was observed in the range of 72.18 and 83.54 (Table 3). This refers to highly bright β-glucan gum pellets and have a great prospective to be used in various food products especially it has great potential in the light colored or transparent types of food products since their incorporation would avoid off or darker colors to these food products. All of the samples differed significantly for each of the parameter of L*A*B* color space. Acidic extraction process yielded aproduct that was higher in b* value with a slightly yellowish outlook such gum pellets are very suitable for incorporation in dips, soups and sauces.

Table 3. L* a* b* color profile of β-Glucan

	L*	a*	b*
M1	72.18 [c]	9.36 [b]	38.65 [a]
M2	79.21 [b]	12.86 [a]	30.53 [b]
M3	83.54 [a]	7.26 c	23.45 [c]
CV (%)	2.36	3.12	1.65
LSD	4.06	1.354	0.726

M1=Acidic Extraction Method; M2=Alkaline Extraction Method; M3=Enzymatic Extraction Method.

2.7. Fourier Transform Infra Red (FTIR) Analysis

FTIR (Fourier transform infra red) is an ideal tool to study the structural features of extracted β-glucan gum pellets, samples were run through Fourier transform infra red spectrophotometer. The FTIR (Tensor 27, Burkus, Germany) used was equipped with MIR-TGS detector and OPUS 5.5 software. A total of 128 scans at a resolution of 8 cm−1 used to observe infra red spectra. The presence of polysaccharides in the present study was confirmed with the appearance of absorption bands in region 1000-1200 cm^{-1}. The absorption peaks at 1074 as revealed by IR spectrum of β-glucan gum pellets (Figure. 5) was attributed to stretching of (CC) (CO) groups and represent presence of glucopyranose. The linear structure with (1→3) linkage was confirmed by the presence of absorption peaks at 1074 cm^{-1} along with peaks in the region of 1156-65.

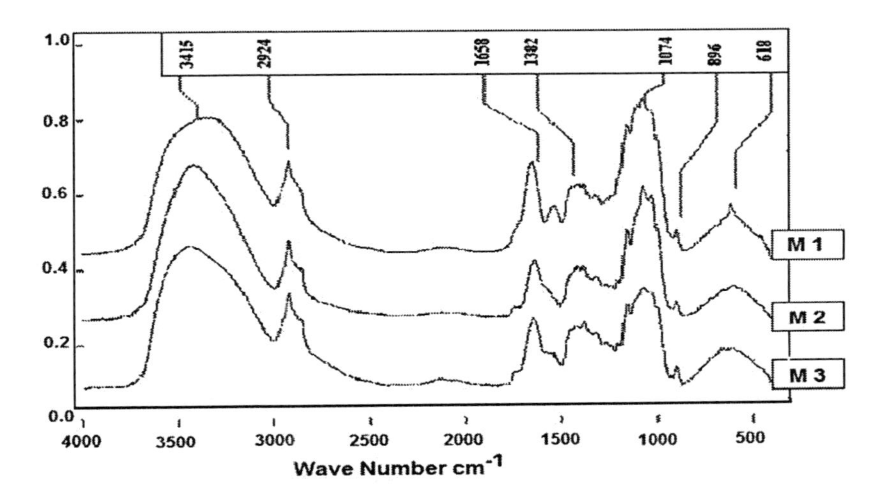

Figure 5. FTIR spectra of β-glucan.

The presence of β_linked glycosidic bonds was substantiate by the appearance of absorption peak at 896 cm^{-1} , There was no peak appeared in region of 840 cm^{-1} that is an indication that extracted β-glucan gum pellets was devoid of α- linked glycosidic bonds . The presence of protein and other nitrogenous compounds were validated by the presence of absorption bands at 1566, 1658cm^{-1} . These bands appeared due to stretching of CN and NH groups of proteins and are indicator of amide I and amide II. The absorption in the region of 1325 and 3415 cm^{-1} was due to stretching of hydroxyl group of water. As all of the samples of gum pellets was chemically characterized with the presence of crude fat that was also observed in FTIR spectra in form of small peaks that was observed at 2855 cm^{-1}.

2.8. Glucose and Lipo Protein Profile

To conduct biological study (Efficacy study), young Albino Rats of the strain of Sprague Dawley were obtained from the National Institute of Health

(NIH), Islamabad. All of the rats was male with almost same age (180±2 days). All of the rats were divided randomly into 6 clusters: First cluster was designated as control and fed on normal control diet with no dietary fiber in their daily diet. To the 2nd cluster a diet was given having β-glucan @ 1 %. For 3rd, 4th, 5th and 6th cluster of rats the feed contained β-glucan @ 2, 3, 4 and 5% level respectively. β-glucan gum pellets having a composition of SDF 71.3±2.73 and IDF 13.6±1.05 was used in biological assay experiment. The housing conditions for rats was keep constant with the maintenance of temperature at 25±2 °C, relative humidity of 60±2%, and a 12:12-hour light-dark cycle with no natural light. During first week of acclimatization, the rats were fed on a standard rat chow in meal. After acclimatization phase, the rats were fed on the experimental diets for 4 weeks. During this period water supply was ad libitum and the animals was free of stress. The diets provided to animals during experimentation period was iso-proteinaceous. Starch, fats, minerals and vitamins were was also added in equal amounts as per requirements of the animals. Only dietary fiber in shape of oat β-glucan gum pellets was in variable amounts. The necessary data for feed intake, weight of each individual rat, feaces and remaining feed residues were recorded once in a day (data not shown).

At the end of the experiment, the rats were decapitated and their blood was collected, centrifuged and serum was preserved at a temperature of 4 °C. Plasma glucose was determined by the method of Thomas and Labor (1992)., Triglycerides, Total cholesterol and HDL were determined using commercial kits from Human (Wiesbaden, Germany). LDL was calculated as the difference between the serum cholesterol value and HDL.

A decrease in serum glucose was observed that vary significantly ($p \leq 0.05$) in all treatments reduced by feeding diets containing β-glucan at various levels. This decline in serum glucose was linear with higher level of decrement was observed as the level of β-glucan was increased in the diet. A positive correlation of between decrease in serum glucose and SDF was found ($r = 0.87$) similar positive correlation was observed for TDF ($r = 0.81$) contents in gum pellets. A decrease of 18.55% as compared to control was recorded when β-glucan gum pellets was incorporated at rate of 5% (Figure 6). This decline in serum glucose was attributed to delayed glucose absorption in small intestine due to increased intestinal viscosity of β-glucan containing diet.

β-glucan containing diets also resulted in serum lipids (triglycerides, LDL and total serum cholesterol) decline. A significant ($p \leq 0.05$) variation was observed in these parameters when diets containing β-glucan was fed to test animals. Comparing all clusters of rats, group of rats that was administrated on diets containing 5% β-glucan showed lowest ($p \leq 0.05$) total serum cholesterol (Figure 7), triglycerides (Figure 8) and LDL-cholesterol (Figure 9).

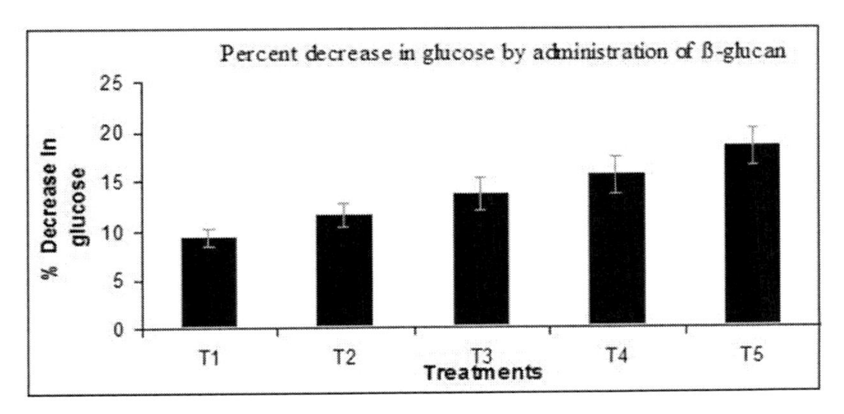

Figure 6. Percent decrease in glucose by administration of β-glucan gum pellets: T_1 =1%, T_2 =2%, T_3 =3%, T_4=4%, T_5 =5%.

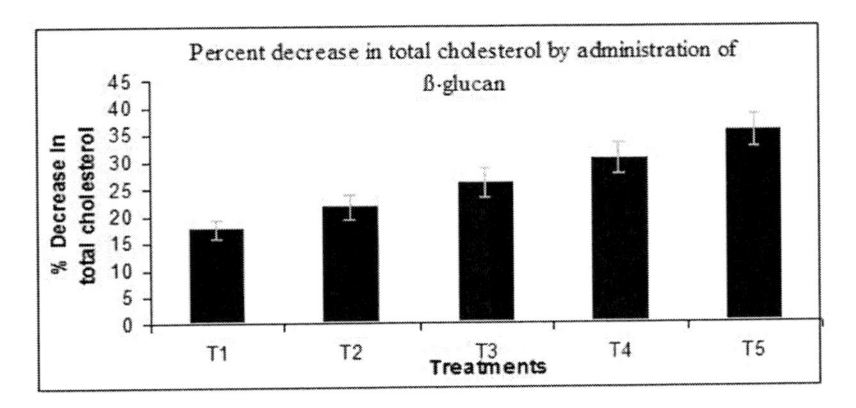

Figure 7.Percent decrease in cholesterol by administration of β-glucan gum pellets: T_1 =1%, T_2 =2%, T_3 =3%, T_4=4%, T_5 =5%.

Correlation study indicated a negative correlation (r = -0.77) between SDF of gum pellet and total serum cholesterol. Similarly, negative correlation was observed for LDL (r = -0.89) and triglycerides (r = -0.74) with SDF of gum pellets. Extracted β-glucan impede the absorption of lipids and cholesterol with a control in micellar solubility. This was further supported by presence of SDF that either tends to reduce reabsorption of bile acids or improve their fecal excretion.

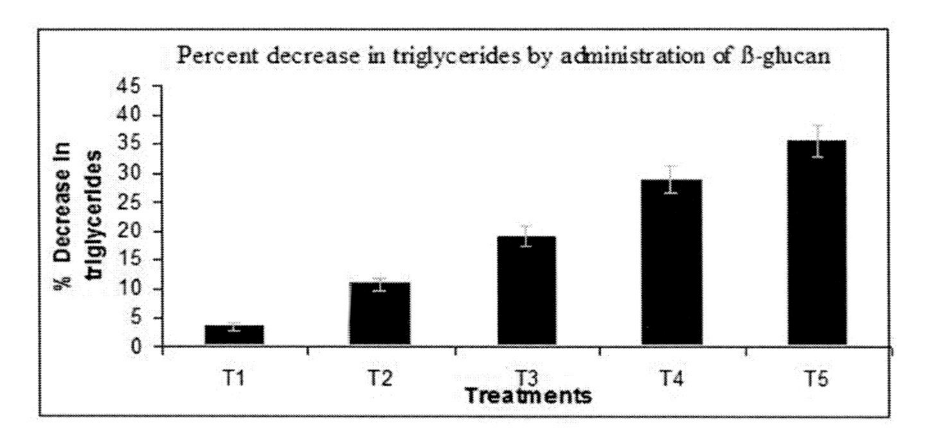

Figure 8. Percent decrease in triglycerides by administration of β-glucan gum pellets T_1 =1%, T_2 =2%, T_3 =3%, T_4=4%, T_5 =5%.

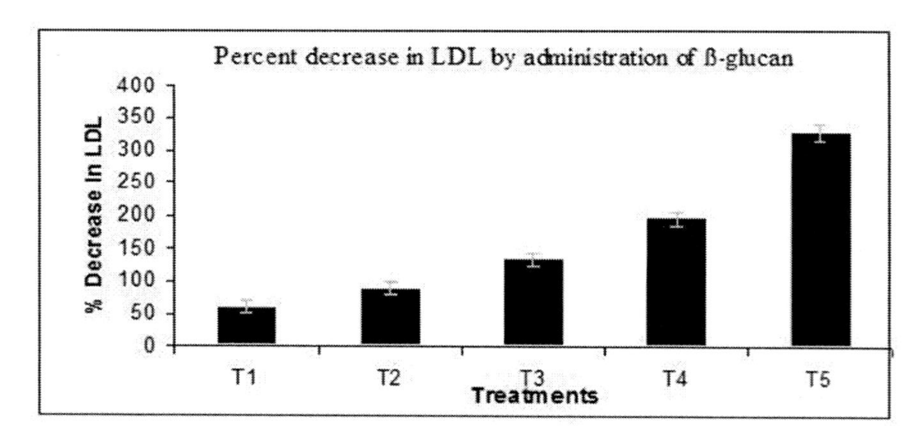

Figure 9. Percent decrease in LDL by administration of β-glucan gum pellets T_1 =1%, T_2 =2%, T_3 =3%, T_4=4%, T_5 =5%.

As a feed back mechanism, their will be more utilization of indigenous serum cholesterol and lipoproteins products for the production of bile acid from cholesterol.

The major discovery in this study pointed out that extracted oat β-glucan gum pellet when incorporated in test diet @ 5% resulted a reduction of 329% LDL as compared to control. Previous work of Brown *et al.*, (1999) indicated that one gram of soluble fiber from oats can produce change in LDL cholesterol of -1.23 mg/dl.

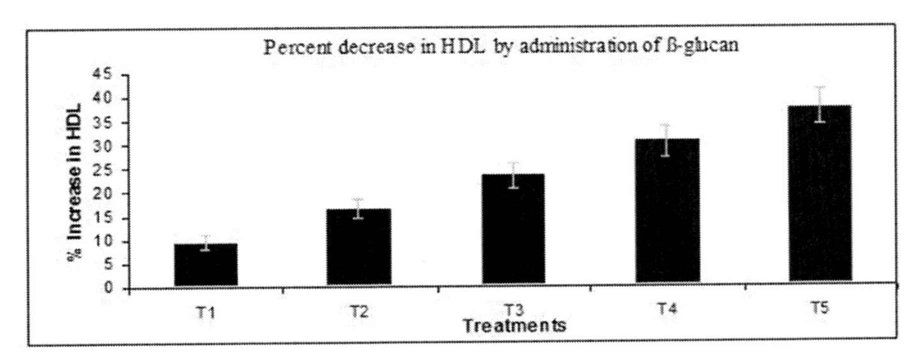

Figure 10. Percent increase in HDL by administration of β-glucan gum pelletsT_1=1%, T_2=2%, T_3=3%, T_4=4%, T_5=5%.

Therefore, our work extends these studies by showing that bioactive β-glucan gum pellets when used at level of 5% can reduce total cholesterol and triglycerides up to 35.68% and 35.55%, respectively. At the same time feeding the rats on β-glucan diet resulted in an increase in HDL by 37.74% as compare to control (Figure 10).

Biological active oat β-glucan gum pellets in the present study showed a great potential in reducing serum cholesterol due to higher SDF and TDF content and balanced composition. These data strongly suggest that incorporation of oat β-glucan gum pellets in the normal diet can be nutritionally important with respect to improvement in lipoprotein profile. Because most of the cholesterol absorbed by mono gastric animals and humans is recirculating endogenous biliary cholesterol, ingestion of β-glucan in these models will reduce cholesterol absorption that would tend to lower serum cholesterol regardless of the amount of dietary cholesterol taken in.

CONCLUSION

Selection of extraction process is vital for specific end use of β-glucan because most of the physicochemical properties of β-glucan are influenced by the choice of extraction method. Among all tested extraction techniques enzymatic extraction process showed a great prospective for industrial applications. As this extraction technique not only removed much of the impurities but was also yielded highest amount of yield and recovery. Small amounts of lipids and fair amount of mineral was also extracted during extraction in all methods. Functional properties like viscosity, foaming capacity, foam stability and water binding capacity was appeared to a high level in all β-glucan samples irrespective the method it was extracted, this make obvious for its candidacy as a new ingredient for industrial usage. Correlation studied indicated a positive correlation between oat β-glucan gum pellet and soluble fiber that in turn provide beneficial health effects. A concomitant decline in glucose, triglycerides, total cholesterol, and LDL can be achieved by increasing dose of β-glucan up to 5%. Negative correlation between SDF and TDF with lipoprotein profile indicated that gum pellets extracted in present study under specific condition have a capacity to reduce total cholesterol, triglycerides and LDL. Extracted β-glucan gum pellets was especially very effective in lowering of LDL cholesterol.

REFERENCES

AACC. 2000. Approved Methods of American Association of Cereal Chemists. *The American Association of Cereal Chemists*, Inc. St. Paul. Minnesota, USA.

Ahmad, A., F.M. Anjum, T. Zahoor, H. Nawaz, A. Din. (2009). Physicochemical and functional properties of barley β-glucan as affected by different extraction procedures. *Int. J. Food Sci. Technol.*, 44(1): 181-187.

Anderson, J.W. (1983). Plant fibre and blood pressure. *Ann. Int. Med.*, 98: 842-846.

Anderson, J.W., (1990). Dietary fibre and human health. *Hort Sci.*, 25: 1488-1495.

AOAC. (1990). Official Methods of Analysis. *The Association of the Official Analytical Chemists*. 15th ed. Arlington Virginia, USA.

Behall, K.M., D.J. Scholfield and J. Hallfrisch. (1997). Effect of beta-glucan level in oat fiber extracts on blood lipids in men and women. *J. Am. Coll. Nutr.*, 16: 46-51.

Bhatty, R.S. (1993). Extraction and enrichment of (1→3) (1→4)- β -D-glucan from barley and oat brans. *Cereal Chem.*, 70 (1): 73-77.

Bhatty, R.S. (1995). Laboratory and pilot plant extraction and purification of β-glucans from hull-less barley and oat bran. *J. Cereal Sci.*, 22: 163-170.

Bingham, S.A., N.E. Day, R. Luben, P. Ferrari, N. Slimani, T. Norat, F. Clavel-Chapelon, E. Kesse, A. Nieters, H. Boeing, A. Tjønneland, K. Overvad, C. Martinez, M. Dorrensoro, C.A. Gonzalez, T.J. Key, A. Trichopoulou, A. Naska, P. Vineis, R. Tumino, V. Krogh, H.B. Bueno-de-

Mesquita, P.H.M. Peeters, G. Berglung, G. Hallmans, E. Lund, G. Skele, R. Kaaks and E. Riboli. (2003). Dietary fiber in food and protection against colorectal cancer in the European Prospective Investigation into Cancer and Nutrition (EPIC): an observational study. *Lancet*, 361: 1496-1501.

Bourdon, I., W. Yokoyama, P. Davis, C. Hodson, R. Backus, D. Richter, B. Knuckles and B.O. Schneeman. (1999). Postprandial lipid, glucose, insulin, and cholecystokinin responses in men fed barley pasta enriched with ß-glucan. *Am. J. Clin. Nutr.*, 69(1): 55-63.

Brennan, C.S. and C.M. Tudorica. (2003). The role of carbohydrates and nonstarch polysaccharides in the regulation of postprandial glucose and insulin responses in cereal foods. *J. Nutraceut. Funct. Med. Foods*, 4: 49-55.

Brown, L., B. Rosner, W. Willet and F.M. Sacks. (1999). Cholesterol lowering effects of dietary fiber: A meta analysis. *Am. J. Clin. Nutr.*, 69 (1):30- 42.

Burkus, Z. and F. Temelli. (2000). Stabilization of emulsions and foams using barley beta-glucan. *Food Res. Int.*, 33: 27-33.

Burkus, Z. and F. Temelli. (2005). Rheological properties of barley β-glucan. *Carbohydr. Polym.*, 59: 459–465.

Cavallero, A., S. Empilli, F. Brighenti and A.M. Stanca. (2002). High (1→3, 1→4) - β-glucan barley fractions in bread making and their effects on human glycemic response. *J. Cereal Sci.*, 36: 59-66.

Crittenden, R., S. Karppinen and S. Ojanen, (2002). In vitro fermentation of cereal dietary carbohydrates by probiotic and intestinal bacteria. *J. Sci. Food Agric.*, 82: 781-789.

Dawkins, N.L. and I.A. Nnanna. (1993). Oat Gum and β-Glucan Extraction from Oat Bran and Rolled Oats: Temperature and pH Effects. *J. Food Sci.*, 58(3): 562–566.

FDA. (1997). Food labeling: health claims; oats and coronary heart disease. Rules and Regulations, *Federal Register,* 62: 3584–3601.

Flander, L., M. Salmenkallio-Marttila, T. Suortti and K. Autio. (2007). Optimization of ingredients and baking process for improved whole meal oat bread quality. *LWT,* 40: 860–870.

Hallfrisch, J., D.J. Schofield and K.M. Behall. (2003). Physiological responses of men and women to barley and oat extracts. (NurimX). II . Comparison of glucose and insulin responses. *Cereal Chem.*, 80, 80-83.

Hill, M.J. (1998). Cereals, dietary fiber and cancer. *Nutr. Res.*, 18: 653-659.

Jenkins, A.L., D.J. Jenkins, U. Zdravkovic, P. Wursch and V. Vuksan. (2002). Depression of the glycemic index by high levels of β-glucan fiber in two functional foods tested in type 2 diabetes. *Eur. J. Clin. Nutr.*, 56: 622-628.

Kahlon, T.S. and F.I. Chow. (1997). Hypocholesterolemic effects of oat, rice, and barley dietary fibers and fractions. *Cereal Foods World*, 42: 86-92.

Kahlon, T.S., F.I Chow, B.E Knuckles and M.M. Chiu. (1993). Cholesterol lowering in hamsters of ß-glucan-enriched barley fraction, dehulled whole barley, rice bran and oat bran and their combinations. *Cereal Chem.*, 70: 435-440.

Keogh, G.F., G.J. Cooper, T.B. Mulvey, B.H. McArdle, G.D. Coles, J.A. Monro and S.D. Poppitt. (2003). Randomized controlled crossover study of the effect of a highly β-glucan–enriched barley on cardiovascular disease risk factors in mildly hypercholesterolemic men. *Am. J. Clin. Nutr.,* 78: 711–718.

Lazaridou, A., C.G. Biliaderis. (2007). Molecular aspects of cereal β-glucan functionality: Physical properties, technological applications and physiological effects. *J. Cereal Sci.,* 46 (2): 101–118.

Li, J., T. Kaneko, L.Q. Qin, J. Wang and Y. Wang. (2003). Effects of barley intake on glucose tolerance, lipid metabolism, and bowel function in women. *Nutrition,* 19 (11-12): 926-929.

M. Irakli, C.G. Biliaderis, M.S. Izydorczyk, I.N. Papadoyannis. (2004). Isolation, structural features and rheological properties of water-extractable β-glucans from different Greek barley cultivars. *J. Sci. Food Agric.*, 84(10): 1170-1178.

Maki, K.C., M.H. Davidson, K.A. Ingram, P.E. Veith, M. Bell and E. Gugger. (2003). Lipid responses to consumption of a beta-glucan containing ready-to eat cereal in children and adolescents with mild-to moderate primary hypercholesterolemia. *Nutr. Res.*, 23: 1527-1535.

Newman, R.K., C.F. Klopfenstein, C.W. Newman, N. Guritno and P.J. Hofer. (1992). Comparison of the cholesterol lowering properties of whole barley, oat bran and wheat red dog in chicks and rats. *Cereal Chem.*, 69: 240-244.

Odes, H.S., H. Lazovski, I. Stern, Z. Madar. (1993). Double-blind trial of a high dietary fiber, mixed grain cereal in patients with chronic constipation and hyperlipidemia. *Nutr. Res.*, 13(9): 979-985.

Pick, M.E., Z.J. Hawrysh, M.I. Gee, E. Toth, M.L. Garg and R.T. Hardin, (1996). Oat bran concentrate bread products improve long-term control of diabetes: A pilot study. *J. Am. Diet. Assoc.,* 96: 1254-1261.

Sier, C.F.M., K.A. Gelderman, F.A. Prins and A. Gorter. (2004). Beta-glucan enhanced killing of renal cell carcinoma micrometastases by monoclonal antibody C250 directed complement activation. *Int. J. Cancer*, 109: 900-908.

Temelli F. (1997). Extraction and functional properties of barley β-glucan as affected by temperature and pH. *J. Food Sci.*, 62: 1192–1201.

Temelli, F. (2001). Potential Food Applications of Barley Glucan Concentrate, Alberta Barley Commission, Barley County, Canada.

Thomas, L. and U. Labor. (1992). Enzymateischer kinetischer colorimetrischer test (GOD-PAP). Biocon Diagnostik, Hecke 8, 34516 Vohl/Manenhagen, Germany. Diagnose. 4: 169.

Tungland, B.C. (2003). Fructooligosaccharides and other fructans: Structures and occurrence, production, regulatory aspects, food applications and nutritional health significance. *ACS Symposium Series*, 849: 135-152.

Valle-Jones, J.C. (1985). An open study of oat bran meal biscuits ('Lejfibre') in the treatment of constipation in the elderly. *Curr. Med. Res. Opin.*, 9(10): 716-720.

Wisker, E., W. Feldheim, Y. Pomeranz and F. Meuser. (1985). Dietary fiber in cereals. In: *Advances in Cereal Science and Technology*, (Ed. Pomeranz, Y). Vol. VII. Am. Assoc. Cereal Chem., St. Paul, Minnesota. 169-238.

Wood, P.J. (2002). Relationships between solution properties of cereal β-glucans and physiological effects. *A review. Trends Food Sci. Technol.*, 13: 313–320.

Wood, P.J., (1993). Physicochemical characteristics and physiological properties of oat (103), (104)-ß-Dglucan. In: Wood, P.J. (Ed.). *Oat Bran. Am. Assoc. Cereal Chem.*, pp: 87-93.

Wood, P.J., D. Paton, I.R. Siddiqui, (1977). Determination of β-glucan in oats and barley. *Cereal Chem.*, 54: 524-533.

Wood, P.J., I.R. Siddiqui and D. Paton. (1978). Extraction of high viscosity gums from oats. *Cereal Chem.*, 54 (3): 524-533.

Wood, P.J., J.T. Braaten, W.S. Fraser, D. Riedel and L.M. Poste. (1990). Comparisons of viscous properties of oat and guar gum and the effects of these and oat bran on glycemic index. *J. Agric. Food Chem.*, 38: 753-757.

Yang, J.L., Y.H. Kim, H.S. Lee, M.S. Lee and Y.K. Moon. (2003). Barley ß-glucan lowers serum cholesterol based on the up-regulation of cholesterol 7-"- hydroxylase activity and mRNA abundance in cholesterol fed rats. *J. Nutr. Sci.*, 49: 381-387.

Yao, N., J. Jannink, P.J. White. (2007). Molecular Weight Distribution of (1→3)(1→4)-β-Glucan Affects Pasting Properties of Flour from Oat Lines with High and Typical Amounts of β-Glucan. *Cereal Chem.*, 84(5) 471–479.

In: Oats: Cultivation, Uses and Health Effects ISBN 978-1-61324-277-3
Editor: D. L. Murphy, pp. 147-157

Chapter 5

OAT GUM FROM OAT GENOTYPES HARVESTED UNDER DIFFERENT IRRIGATION CONDITIONS: EXTRACTION, CHARACTERIZATION AND GELLING CAPABILITY

Naivi C. Ramos-Chavira[a], Elizabeth Carvajal-Millan*[,c], Alma C. Campa-Mada[c], Agustin Rascon-Chu[d], Victor M. Santana-Rodriguez[a], Juan J. Salmerón-Zamora[b] and Armando Quintero-Ramos[a]

[a]Faculty of Chemistry
[b]Faculty of Agro-technological Sciences
Autonomous University of Chihuahua, 31125 Chihuahua, México
[c]CTAOA, Laboratory of Biopolymers
[d]CTAOV, Laboratory of Biotechnology
Research Center for Food and Development, CIAD,
A. C., Carretera a La Victoria Km 0.6
Hermosillo, Mexico

---

* Corresponding author: Tel/Fax: +52 662 2892400; E-mail address: ecarvajal@ciad.mx

ABSTRACT

Research on oat (*Avena sativa)* has been intensified in the last years as it has been reported to reduce serum cholesterol levels and attenuate postprandial blood glucose and insulin responses, which has been related to the presence of β-glucan. *A. sativa* is extensively planted as a forage crop in Northern Mexico, where rainfall has an erratic distribution resulting in smaller oat grains, which failed to meet the requirements of the market. In this regard, the extraction of a β-glucan-enriched oat gum from drought harvested *A. sativa* seeds could be an interesting alternative for this agricultural product to still represent value to producers. Mexican oat genotypes are important animal feed resources and have been studied on the basis of forage yield and nutritional value. Nevertheless, to our knowledge, studies on the extraction and characterization of oat gum from *A. sativa* Mexican genotypes harvested under different irrigation conditions have not been reported elsewhere. This chapter has been focused on the extraction and evaluation of the physicochemical and gelling properties of β-glucan enriched oat gums from three oat genotypes (Karma, Cevamex and Cuauhtémoc) under two irrigation conditions: rainfed farming (RF) and irrigated crop (IC). Oat gums from RF Karma, Cevamex and Cuauhtémoc presented a β-glucan content of 62, 56 and 54 % (w/w), while lower values where found for IC genotypes (61, 50, 47 % w/w, respectively). The intrinsic viscosity $[\eta]$ and viscosimetric molecular weight (Mv) values of oat gums extracted from RF samples were higher than those of the corresponding IC genotypes. The oat gums formed physical gels (10% w/v in water) after heating at 75°C for 1 h and cooling at 25°C for 2 h. Large deformation mechanical tests (compression mode) revealed an increase in hardness of oat gums gels with increasing $[\eta]$ and Mv.

Keywords: Oat glucans, gelation, dietary fibre, antioxidant.

INTRODUCTION

β-D-glucan is a linear, unbranched polysaccharide containing a single type of monosaccharide, i. e. β-D-glucose. The β-D-glucose has two types of linkage, one is (1→4)-O-linked β-D-glucopyranosyl unit (~70-72%) and the other is (1→3)-O-linked β-D-glucopyranosyl (~28-30%) unit. The structure features the presence of consecutive (1→4)-linked β-D-glucose (mostly 2 or 3, and sometimes up to 14 blocks) with blocks that are separated by single

(1→3)–linkage. The (1→3) (1→4) β-D-glucans are cell wall polysaccharides of cereal endosperm and aleurone cells. The content of β-D-glucans in cereals follows the order of barley 3-11%, oat 3.2-6-8%, rye 1-2% and wheat < 1% (Cui, 2001). β-glucans exhibit interesting physicochemical properties, especially gelling capability, leading to their extensive use in the food industry as encapsulation agents for flavors, colors, antioxidants, microorganisms and others (Johansson et al., 1993). Moreover, they also exhibit biological activities associated with medical, pharmaceutical and cosmetic applications (Laroche and Michaud, 2007). Functional properties of the gums depend on two factors: The intrinsic molecular characteristics including conformation, molecular weight, degree of branching and ionization, and extrinsic factors such as pH, concentration, ionic strength, and temperature which may also influence the flow behavior. Each gum has specific physical and chemical characteristics that cannot easily be replaced with the use of another polysaccharide (Badui, 1999). In the last two decades, β-glucans have attracted much attention due to their acceptance as functional, bioactive ingredients. β-glucans have also been associated with human health benefits in prevention and treatment of coronary health disease and diabetes, reduction of plasma cholesterol and a better control of postpandrial blood serum glucose and insulin responses (Wood et.al., 1994; Wood, 2007).

Oat is a grain with a high content of β-glucans. States located in Northern Mexico are the largest producers of oat in our country. However, this grain is mostly sold without being processed for a low price. Extraction of β-glucans from oat represents an attractive option, since in recent years it has been worldwide reported an increase in their use in both food and pharmaceutical industry. Extraction and evaluation of the physicochemical and gelling properties of β-glucan enriched oat gums from three oat genotypes (Karma, Cevamex and Cuauhtémoc) under two irrigation conditions: rainfed farming (RF) and irrigated crop (IC) were the main objectives for this research.

Experimental

Materials

Oat genotypes used in this project (Cevamex RF, Cevamex IC, Cuauhtémoc RF, Cuauhtémoc IC, Karma RF, Karma IC) were provided by the National Institute for Investigation in Forestry, Agriculture and Animal Production in Mexico (INIFAP). The samples of whole grain oat of each genotype were received in polyethylene bags of 5 kg each and remained stored at 4°C. Grain husk was removed, followed by a milling process where the oat grain is ground into flour (0.84 mm particle size) using a M20 Universal Mill (IKA®, Werke Staufen, Germany).

Oat Gum Extraction

Oat gum was water extracted from milled seeds (1 kg/ 3 L) for 15 min at 25°C. The water extract was then centrifugated (12096 *g*, 20°C, 15 min) and supernatant recovered. Supernatant was precipitated in 65% ethanol treated for 4 h at 4°C. Precipitate was recovered and dried by solvent exchange (80 % (v/v) ethanol, absolute ethanol and acetone) to give oat gum (Ramos-Chavira et al., 2009).

Chemical Analysis

The β-glucan content was determined as previously reported (McClearly et al., 1985). Neutral sugars content was according to Carvajal-Millan et al. (2007). Samples were analyzed by high performance liquid chromatography (HPLC) using a Supelcogel Pb column (300 x 7.8 mm; Supelco, Inc., Bellefont, PA) eluted with water (filtered 0.2 μm, Whatman) at 0.6 mL/min and 80 °C. Inositol was used as internal standard. A refractive index detector Star 9040 (Varian, St. Helens, Australia) was used. Ferulic acid content was determined by RP-HPLC as reported by Vansteenkiste et al. (2004). Detection was by UV absorbance at 280 nm. A Varian 9012 photodiode array detector (Varian, St. Helens, Australia) was used (Vansteenkiste et al., 2004). Protein content was determined according to Bradford (1976). Ash content was according to the AACC approved method (AACC, 2000). Results were reported on a dry weight basis (d.b.).

Intrinsic Viscosity and Viscosimetric Molecular Weight

Specific viscosity (ηsp) was measured by registering oat gum solutions flow time in an Ubbelohde capillary viscometer at 25 ± 0.1 °C, immersed in a temperature controlled bath. The intrinsic viscosity ([η]) was estimated from relative viscosity measurements, ηrel, of oat gum solutions by extrapolation of Kraemer and Mead and Fouss curves to "zero" concentration (Mead and Fouss, 1942; Kraemer, 1983). The viscosimetric molecular weight ($M\nu$) was calculated from the Mark–Houwink relationship.

Gelation

Oat gum solutions at 10% (w/v) were prepared by gentle stirring samples in double distilled water at 75°C until complete solubilization of the material. Gels were allowed to form for 2 h at 25°C (Lazaridou et al., 2003).

Gel Hardness

The hardness of 5, 8 and 10% (w/v) oat gum gels, freshly made (2 h) in 6 mL glass flasks of 30 mm height and 25 mm internal diameter was analyzed with a TA.XT2i Texture Analyzer (RHEO Stable Micro Systems, Haslemere, England). The gels were deformed by compression at a constant speed of 1.0 mm/s to a distance of 4 mm from the gel surface using a cylindrical plunger (diameter 15 mm). The peak height at 4 mm compression was called gel hardness (Carvajal-Millan et al., 2005).

RESULTS AND DISCUSSION

Oat gums yield for the different oat genotypes ranged from 2.2 up to 3.2% (w oat gum/w oat seeds), which means that around 60% of the *β*-glucan initially present in the oat seed was recovered (Table 1). According to Beer et al. (1996) 60-65% of total *β*-glucan can be extracted from oat seeds by water extraction at 90°C, which are similar conditions to those used in the present research.

Chemical composition of oat gums is shown in Table 2. The extraction procedure adopted in the present study provided an oat gum with *β*-glucan contents from 47 up to 61 (% w/w) for the different oat genotypes. Oat gums with higher content of β-glucans were Karma RF and Karma IC (61% w/w), and the lowest was Cuauhtémoc IC (47% w/w). Starch, arabinose, xylose, protein, ash and ferulic acid were also detected in these oat gums. There were significant differences in protein content between the gums from the three genotypes and the type of irrigation (P<0.05).

Table 1. Yields of oat gum obtained from oat genotypes harvested under different irrigation conditions

Sample	Yield (w oat gum/w oat seeds)
Cevamex RF	3.2
Cevamex IC	3.2
Cuauhtémoc RF	3.2
Cuauhtémoc IC	2.2
Karma RF	2.3
Karma IC	2.2

Table 2. Compositional characterization of different oat gums extracted

Sample	β-Glucan	Protein	Ferulic Acid	Arabinose	Glucose	Xylose
Cervamex RF	56 ± 3 [ab]	5.7 ± 0.1 [b]	0.020 ± 0.001 [c]	0.75 ± 0.01 [b]	35 ± 4 [ab]	1.11 ± 0.02 [b]
Cervamex IC	50 ± 3 [bc]	3.5 ± 0.1 [d]	0.033 ± 0.002 [ab]	0.79 ± 0.01 [ab]	39 ± 5 [ab]	1.12 ± 0.01 [bc]
Cuauhtémoc RF	54 ± 1 [abc]	3.4 ± 0.3 [d]	0.010 ± 0.001 [d]	0.82 ± 0.03 [a]	36 ± 2 [ab]	1.11 ± 0.04 [bc]
Cuauhtémoc IC	47 ± 1 [c]	4.8 ± 0.1 [c]	0.012 ± 0.001 [d]	0.78 ± 0.01 [ab]	44 ± 1 [a]	0.99 ± 0.02 [c]
Karma RF	61 ± 0.2 [a]	4.8 ± 0.1 [c]	0.022 ± 0.001 [a]	0.80 ± 0.01 [a]	31 ± 5 [b]	1.07 ± 0.01 [b]
Karma IC	61 ± 0.5 [a]	6.1 ± 0.1 [a]	0.031 ± 0.002 [b]	0.75 ± 0.01 [b]	29 ± 3 [b]	1. 18 ± 0.02 [a]

Reported values are the averages of three replications.
% (w/100w oat gum dry basis).
Different superscripts in the same column indicate significant differences P < 0.05.

The gum extracted from the genotype Karma IC had the higher protein content and the lowest was the Cuauhtémoc RF. There were significant differences in the ferulic acid content in genotypes, type of irrigation and its corresponding interaction. Ferulic acid content was higher in IC genotypes. The genotype with higher levels of this antioxidant was Cevamex IC and the least amount was found in Cuauhtémoc RF. The highest content of glucose was found in genotype Cuauhtémoc IC. There were not significant differences in the arabinose and xylose content between oat gums genotypes.

There were significant differences between genotypes, irrigation type and their interaction in the intrinsic viscosity and molecular weight. Figure 1 shows that the highest intrinsic viscosity was for Karma RF (133 mL/g) and the lowest for Cuauhtémoc IC (75 mL/g). The highest molecular weight was registered for Karma RF, with a value of 47 590 Da while the lowest value was found for Cuauhtémoc IC (26 436 Da) (Figure 2).

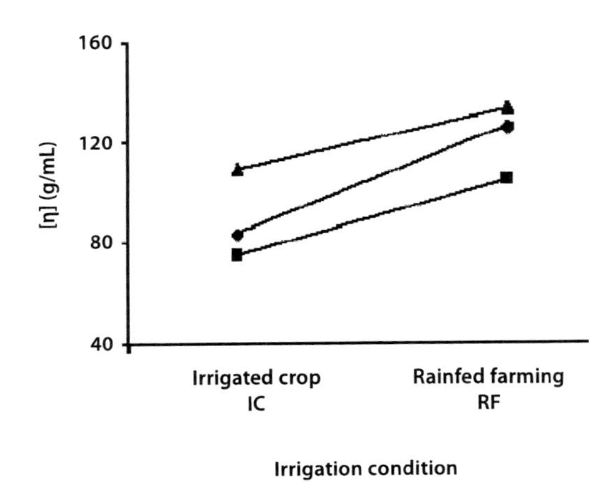

Figure 1. Intrinsic viscosity of oat gums extracted from Cevamex (♦), Cuauhtémoc (■) and Karma (▲) genotypes harvested under different irrigation conditions.

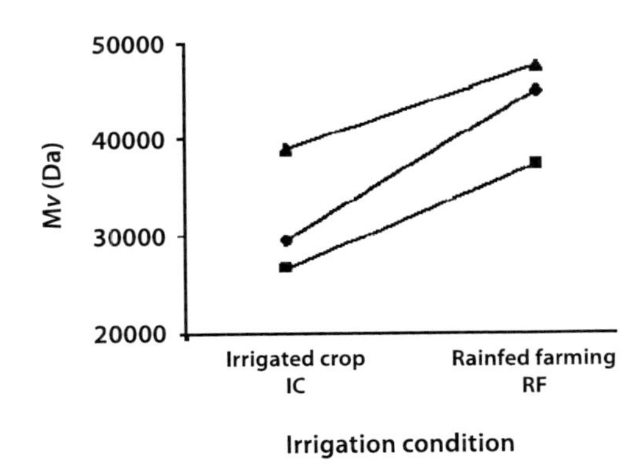

Figure 2. Molecular weight of oat gums extracted from Cevamex (♦), Cuauhtémoc (■) and Karma (▲) genotypes harvested under different irrigation conditions.

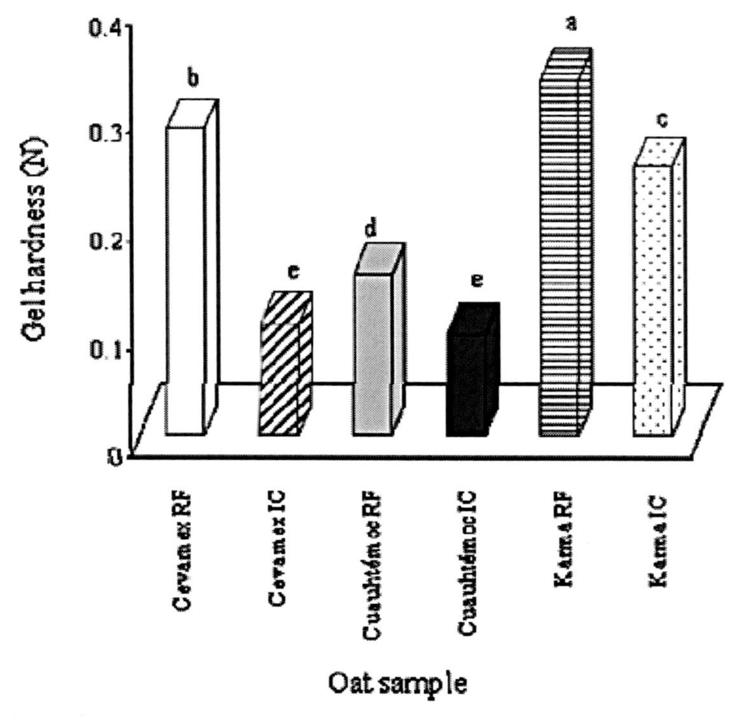

Figure 3. Hardness of oat gum gels at 10% (w/v) from oat genotypes harvested under different irrigation conditions.

Solubilization of oat gums in hot water produced oat gum solutions at 10% (w/v) which formed firm and white gels after 2 h at 25°C. The hardness of oat gum gels is reported in Figure 3.

There were significant differences in the gel hardness between oat gums genotypes. Oat gum from Karma RF formed gels with the highest hardness value (0.33 N) while the lowest value was registered for Cuauhtémoc IC (0.1 N). β-glucan gels belong to the category of physically cross-linked gels whose three-dimensional structure is stabilized mainly by multiple inter and intra chain hydrogen bonds in the junction zones of the polymeric network (Cui, 2001).

From a functional point of view, it has been reported that the gel formation properties of β-glucan increase intestinal transit time (Anderson and Chen, 1986). The gel network that is formed could also slow down the digestion by retarding diffusion and diminishing contact between gastrointestinal enzymes and their substrates (Lazaridou et al., 2003).

CONCLUSION

Oat gums with β-glucan as major component can be recovered from oat genotypes harvested under different irrigation conditions. The oat gums recovered formed physical gels. Rainfed farming conditions increase oat gums intrinsic viscosity and molecular weight improving the gels hardness. Although the amount of ferulic acid present in the different oat genotypes studied is not as great as in other cereals, it is a significant source as an antioxidant. Recuperation of these oat gums from a low value oat grains could represent a commercial advantage face to other gums commonly used in the food industry.

ACKNOWLEDGMENTS

The authors are pleased to acknowledge Jorge Márquez-Escalante, Nora Ponce de León-Renova and Karla G. Martínez-Robinson for their technical assistance.

REFERENCES

AACC. (2000). *Approved Method of the AACC.* Method 26-10. 10th ed., St. Paul, MN: American Association of Cereal Chemists.

Anderson, J.W., Chen, W.J.L. (1986). Cholesterol-lowering properties of oat products. In: F.H. Webster (Ed), *Oats: Chemistry and Technology* (1st.ed, 309-333), St. Paul, MN: American Association of Cereal Chemists.

Badui, S. (1999). Química de los alimentos. México, DF. Limusa.

Beer, M.U., Arrigoni, E., Amado, R. (1996). Extraction of oat gum from oat bran: effects of process on yield, molecular weight distribution, viscosity and (1→3), (1→4)-β-D-glucan content of the gum. *Cereal Chemistry*, 73, 58-62.

Bradford, M. M. (1976). A rapid and sensitive method for the quantification of microgram quantities of protein utilizing the principle of protein-dye binding. *Analytical Biochemistry*, 72, 248-254.

Carvajal-Millan, E., Guigliarelli, B., Belle, V., Rouau, X., Micard, V. (2005). Storage stability of arabinoxylan gels. *Carbohydrate Polymers*, 59, 181-188.

Carvajal-Millan, E., Rascón-Chu, A., Marquez-Escalante, J., Ponce de León, N., Micard, V., Gardea, A. (2007). Maize bran gum: characterization and functional properties. *Carbohydrate Polymers*, 69, 280-285.

Cui, S.W. (2001). *Polysaccharide gums from agricultural products. Processing, structures and functionality*. Lancaster, PA: Technomic Publishing Co. Inc.

Johansson, L., Molin, G., Jeppsson, B., Nobaek, S., Ahrne, S., Bengmark, S. (1993). Administration of different lactobacillus strains in fermented oatmeal soup: in vivo colonization of human intestinal mucosa and effect on the indigenous flora. *Applied and Environmental Microbiology*, 59, 15-20.

Kraemer, E. (1983). Molecular weight of celluloses and cellulose derivates. *Industrial and Engineering Chemistry*, 30, 1200-1203.

Laroche, C., Michaud, P. (2007). New developments and prospective applications for β (1,3) glucans. *Recent Patents on Biotechnology*, 1, 59-73.

Lazaridou,A., Biliaderis, C.G., Izydorczyk, M.S. (2003). Molecular size effects on rheological properties of oat β-glucans in solutions and gels. Food Hydrocolloids, 17, 693-712.

McClearly, B.V., Glennie-Holmes, M. (1985). Enzymic quantification of (1→3), (1→4)- β-D-glucan in barley and malt. *Journal of the Institute of Brewing*, 91, 285-295.

Mead, D., Fouss, R. (1942). Viscosities of solutions of polyvinyl Chloride. *Journal of the American Chemical Society*, 64, 277-282.

Ramos-Chavira, N., Carvajal-Millan, E., Marquez-Escalante, J., Santana-Rodriguez, V., Rascon-Chu, A., Salmerón-Zamora, J. (2009). Characterization and functional properties of an oat gum extracted from a drought harvested A. sativa. *Journal of Food Science and Biotechnology,* 18, 900-903.

Vansteenkiste, E., Babot, C., Rouau, X., Micard, V. (2004) Oxidative gelation of feruloylated arabinoxylan as affected by protein. Influence on protein enzymatic hydrolysis. *Food Hydrocolloids*, 18, 557-564.

Wood, P.J., Braaten, J.T., Scott, F.W., Riedel, K.D., Wolynetz M.S., Collins, M.W. (1994). Effect of dose and modification of viscous properties of oat gum on plasma glucose and insulin following an oral glucose load. *British Journal of Nutrition*, 72, 731-743.

Wood, P.J. (2007). Cereal β-glucans in diet and health. *Journal of Cereal Science*, 46, 230-238.

INDEX

C

D

G

H

M

Q

R

S

T